AF773239

Engineersmanship . . .
The Doing of
Engineering Design

Engineersmanship . . . The Doing of Engineering Design

SECOND EDITION

LEE HARRISBERGER

University of Alabama

Brooks/Cole Engineering Division

Monterey, California

Brooks/Cole Engineering Division
A Division of Wadsworth, Inc.

©1982, 1966 by Wadsworth, Inc., Belmont, California 94002.
All rights reserved.
No part of this book may be reproduced,
stored in a retrieval system, or transcribed, in any form or by any means—
electronic, mechanical, photocopying, recording, or otherwise—
without the prior written permission of the publisher,
Brooks/Cole Engineering Division, Monterey, California 93940,
a division of Wadsworth, Inc.

Printed in the United States of America
10 9 8 7 6 5 4 3 2 1

Library of Congress Cataloging in Publication Data
Harrisberger, Lee.
Engineersmanship—: the doing of
engineering design.

Includes index.
1. Engineering design. I. Title.
TA174.H3 1981 620'.00425 81-6181
ISBN 0-8185-0441-2 AACR2

Production Editor: *Cece Munson*
Designer: *Jamie Sue Brooks*
Illustrations: *Forrest Harrisberger*
Typesetting: *Linda Andrews, Ashland, Oregon*

Preface

In a world of population explosions, nuclear explosions, technological breakthroughs, and everyday miracles, the design engineer is increasingly immersed in the "real life" domain of the world community. Today's projects invariably subject the engineer to a variety of concerns that are unrelated to the traditional technical and analytical processes of engineering. As a result, an engineer becomes deeply involved in such topics as idea-getting, value versus cost, aesthetics, style, appearance, communication, the philosophy of design, professional attitude, and community responsibility.

This book discusses several subjects that are pertinent and enriching to all engineers involved in design activity. It is an attempt to meet the challenge of the statement, "There's more to design than analysis!" A good engineering design is a *complete* engineering design developed by *complete* engineers who are adept at Engineersmanship.

Lee Harrisberger

Contents

Engineersmanship . . .
The Doing of
Engineering Design

1

Engineering . . . Then, Now, and Later

The phenomenal progress of people in mastering the universe can be largely attributed to our ability to adapt forces and laws of nature for personal use. Down through the ages this ability has created the machines and the systems that are responsible for our culture and achievements. This process is engineering. Let's look at how engineering fits into the scheme of things, what design is, and what engineers now have left to conquer through their efforts in engineering design.

ENGINEERING IS . . .

There has been some confusion concerning what engineering is and what design is. There are many definitions for engineering. For example, engineering is:

- the adaptation of scientific discovery to useful purposes.
- the creation of useful devices for the service of people.
- the process of inventing solutions to our needs.
- the solving of problems of a technical nature.

- the conversion of the forces of nature for our purposes.
- the conversion of energy resources into useful work.

Regardless of how you say it, engineering is our effort to control and utilize nature for our own benefit.

ENGINEERING DESIGN IS . . .

When there is a physical problem to be solved, it can be reduced to what is available and what needs to be done, i.e., an input and an output. In between these two situations, a physical device, machine, or process must be installed to convert the available input into the required output. Filling this "space" with a specifically designed system of hardware is engineering design, and it has many facets. It is, for example:

- the act of creating a device, machine, or system which will fill a particular need.
- the development of a mechanism to transform a given input into a particular output.
- the process of devising a scheme to fulfill a particular requirement through the use of some bit of knowledge or an observation of nature.
- the development of a solution to a particular problem.
- the development of some sort of physical apparatus, some sort of hardware (no matter how complex or how simple).
- a problem-solving situation in which the design engineer is trying to satisfy a technological need.
- taking a need and breaking it down into a set of problems; devising solutions to those problems and developing ways to manufacture a product to fill the need.
- the business of inventing and creating ideas which will adapt a scientific discovery to the development of a useful product.
- the recognition of a need, the planning for accomplishment, an application of ideas, the adaptation of natural laws to useful purposes, and a systematic synthesis of cause and effects.
- the process of converting a specified input into a desired output while at the same time acknowledging all of the

environmental constraints that restrict the development of the solution.
- a creative act of selecting, combining, converting, constraining, modifying, manipulating and shaping ideas, scientific facts, and physical laws into a useful product or process.
- demand ideation—creating for a purpose.

In other words, when you are involved in engineering design, you are where the action is!

IT ALL STARTED WAY BACK WHEN . . .

From the time human beings first learned to take advantage of natural phenomena for their own benefit, they have been engaged in engineering. When they discovered that they could prop a stick on a rock and move a boulder that was too heavy to lift any other way, or use a wheel to reduce the friction of dragging something around, or use a water wheel to supplement human labor, they started using nature for their own benefit. The earliest manifestations of our rise from the animal kingdom are examples of our ability to adapt the observations of natural phenomena to our own use. We have never been content to rest on discovery alone. We have always been oriented to adapt our discovery and our knowledge to useful purposes. It is through these efforts that we have achieved our current mastery of the universe.

It is quite obvious that our modern society has been created by engineering and science and that the structure of our society has been profoundly affected by the engineering developments which have occurred throughout history.

IT WASN'T EASY . . .

Progress in mechanization was painfully slow. For thousands of years, people lived in a constant struggle for existence. Their ignorance and fear of nature and the struggle for survival left little time for innovation and education. What progress there was came about initially because people began to band together into groups. Under the leadership of the stronger individuals, some people themselves became tools for the ambitions of others. They became slaves to the leadership of other people. Although slave labor was to sustain societies for thousands of years, it nevertheless deterred

Under the leadership of stronger individuals, some people themselves became tools for the ambitions of others.

the progress of mechanization. There was no need to mechanize when large groups of slaves were available to do whatever task was needed.

Actually many great engineering works were made primarily to serve the superstitions of people and the ambitions of leaders rather than the fundamental needs of humans. Great monuments and "Wonders of the World" were erected to serve the whims of a king or leader. Fantastic hordes of slaves were employed for these tasks. Much early engineering effort was also devoted to the development of weapons. The advancements in metallurgy were attributed to the making of swords and tools for war. Few of the discoveries and engineering applications in those days benefited the masses.

PERFECTION BY ACCIDENT

It is interesting to note that the development of the early mechanical devices and weapons through centuries of trial and error was curiously analogous to the natural evolution of plants and animals. This long-term design process became a sequence of lucky accidents which were passed down from generation to generation, from century to century. Many of the devices that evolved eventually reached a state of optimized perfection.

A typical example of this sort of evolutionary optimization is the development of the bow. Several years ago an experimental stress analysis was made of the classical bow to determine if there was a possibility of optimizing the power of the bow by making it a uniformly stressed beam. It was discovered that the form of the bow had already evolved into a uniformly stressed beam of maximum power potential as is the case in most of the natural evolutions.

Early technical progress in engineering was almost totally accidental. There was no rationale and no fundamental knowledge of nature. People continued to reconstruct these fortuitous accidents methodically and exactly throughout the years in order to preserve the results that were stumbled upon.

There is the story that is told of the development of a particular kind of special sword metal in the early days. As the story goes, the craftsmen who were engaged to make the swords were methodically going through their ritual of heating and tempering and

. . . that some swordmakers improved quality of their swords by thrusting the heated blade into the belly of a slave.

hammering the metal in order to harden it and to improve its grain structure. One day the swordmaker had laid his heated sword blank off to one side, and a little red-haired slave boy urinated on the hot metal (not an unusual impulse for small boys). That particular sword turned out to be far superior to any sword that craftsman had ever made before and he was called upon to make many more of these superior swords. From then on this enterprising craftsman obtained a red-haired slave boy to urinate on all his heated sword blanks. This, no doubt, put quite a premium on red-haired slave boys in that period of time. Other stories go that some swordmakers improved the quality of their swords by thrusting the heated blades into the belly of a slave. This, no doubt, had a moderate tempering effect, although it was hard on slaves.

THE AGES OF ENGINEERING

According to Finch (4),* engineering is one of the oldest of professions, dating back to 4000 B.C. and the fantastic constructions of the Egyptians. The first age of engineering works (although the word *engineering* was unknown) lasted until the fall of the Roman Empire. When civilization plunged into the Dark Ages, all engineering activities essentially disappeared.

The reemergence of engineering activity began to show in the 15th Century during the Renaissance. Two significant events occurred that seem to have had a particular influence on the beginning of the second age of engineering: the invention of the printing press and the decline of slavery and serfdom (cheap labor).

Actually, the present era of engineering could attribute its start to Italy, where there was an acute shortage of labor and the people were not ravaged by the feudal system. It was about this time that one of the first and most complete records of machinery was published by a man named Ramelli in 1588. It has been referred to as one of the first engineering handbooks. This unusual 354-page book contained 195 double-page etched illustrations of a variety of pumps, mills, looms, derricks, elevators, presses, sawmills, and war machines. Most of the machines were made of wood and involved some rather ingenious devices and mechanisms. Many of the mechanisms of that age are the forerunners of our modern gear trains, worm drives, screw drives, crank-and-rocker

*Numbers in parentheses refer to references listed at the end of the chapter.

mechanisms, intermittent mechanisms, reversing clutches, and so on. The machines had hand-carved cog wheels and wooden screws which could run only at very low speeds. The typical power source was a treadmill, turnstyle, windmill, or waterwheel.

The design innovators of the day were the watchmakers, who were responsible for some rather ingenious mechanical innovations. These watchmakers were often called upon as engineering consultants to devise large and unusual machines for special purposes. One example is the record of a pumping machine which was built in Spain in 1588. An Italian clockmaker, Giovanni Turriano, was commissioned to design a machine to raise water 300 feet from the Tagus River to the Alcazar Palace. This fantastic

. . . it was amazing that it should work at all.

machine deserves mention if for no other reason than it was amazing that it should work at all. According to the historical notes, it didn't work long. It certainly is an example of the ingenious and amazingly complex apparatus that was conceived of in those days.

One of the most famous machines of this era was the Marly machine which was built for Louis XIV by the Flemish engineer R. E. Rennequin Sualem in 1683 to supply water for the fountains of Versailles. It was truly one of the mechanical wonders of the day. Costing over $10 million, it contained 250 pumps, provided

533 feet of lift, and was driven by 14 undershot waterwheels, 40 feet in diameter. These waterwheels were connected to an astonishing system of linkage containing over 64,000 feet of iron bars that extended 2,020 feet up the hill from the river to provide three stages of pumping. This monstrous system had 64 pumps mounted at the river level, 79 pumps at 160 feet above the river driven by seven linkages 600 feet long, and 82 pumps located at 245 feet above the river level driven by 13 linkage systems 2,020 feet long. All of these linkage systems were connected by cranks to the 14 waterwheels that provided approximately 1,200 HP to the system. It had an incredibly low efficiency and most of the power was lost to friction and inertia and noise. The clank of the machine was heard for miles. It has been estimated that the annual upkeep was over $60,000 a year. And, most unbelievable of all, this monstrosity was in active service for over 120 years!

The irony of the effort involved in the Marly machine was that its sole purpose was to supply a substantial head of water for the 60-feet-high jets and cascades in the gardens of the palace at Versailles. This was typical of the age. Many of the remarkable engineering developments were carried out to satisfy the whim of a monarch. Such feats had very little application to the needs of the community at large.

The revival of engineering was sustained by the upsurge of development in both Italy and France terminating with the collapse of the French economy during the French Revolution in the late 1700s. Even so, the accomplishments were limited to impressive masonry constructions and water works, since power was still limited to human, animal, and water sources. It was, however, during this time that those who created them began to be referred to as "ingeniators," the origin of the title "engineer."

It can easily be argued that the birth of "modern" engineering occurred with the beginning of the Industrial Revolution in Britain in the late 1700s and early 1800s. Breakthroughs in metal production and steam-power generation spawned an avalanche of mechanical innovations and opportunities for a new wave of engineering activity.

Parallel with an unprecedented opportunity for engineering progress was the emergence of science as a source discipline of new knowledge (rather than as a philosophy) and the development of the science of mathematics in France. The merging of these three disciplines as allies converted engineering from an empirical trade and craft into a rational discipline of applied science, analysis, and design. The discipline of engineering had become equipped to create technological innovations. All it needed was an affluent, ambitious customer and America was it!

The unprecedented exponential growth in the application of scientific discoveries and the technical sophistication of the entire social structure is a result of this partnership. Now engineering and engineers seem to be as basic to survival as food and water.

THE PATTERN OF EVOLUTION

Engineering progress has been totally dependent upon the evolution of power, materials, and tools. Each has been dependent on the others. The compounding interaction of need versus power versus material versus tools creates a curious and typical pattern of design evolution. The introductory model of a new engineering concept is, in retrospect, a crude and limited prototype of the optimum design. Nevertheless, it still represents the best engineering capability that the technology and economy can bring to bear. The limitations of this prototype set off a variety of technological developments bent on making it possible to improve the device. "Necessity is the mother of invention."

There is a typical pattern in the development of an engineering innovation. For a considerable period of time there is little improvement in the product. If the product has a great market potential and widespread need, it will justify its own research and development program to overcome the obstacles of its improvement. Otherwise, its evolution will depend on the adaption of technical innovations which evolve during the solution of other, more pressing, engineering problems. These studies eventually set off a sequence of improvements in the product, a chain reaction in modification. There comes a time, however, in the evolution of every product, when effort needed to improve outweighs the gains in product usefulness. It has been optimized. It has reached a plateau of development: it is a highly reliable, useful product that's about as good as it can get.

The washing machine is a typical example of this type of product evolution. For many years the hand-powered, wooden-tub washer with hand-powered wringer was the only home-laundry device. The advent of electricity provided electric-motor drives to the washer; the wooden tub gave way to galvanized metal, then porcelain; the wooden rollers to soft rubber; the wooden paddles

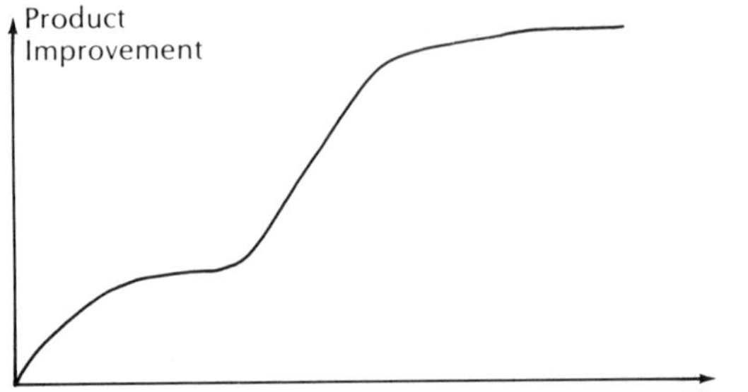

to a molded plastic agitator. By the 1930's, the "wringer washer" had reached a plateau of development. Then came the "break-through" via the automatic washer which eliminated the tubs, the wringer, and the labor. Soon the automatic clothes dryer came along as an adjunct to the home laundry, and again, a plateau has been arrived at in the development of the washing machine. Virtually all the inherent inconveniences of the "wet-washing" process have been eliminated in the evolution of the automatic washer-dryer home-laundry system. Note, however, that the concept of cleaning by agitating clothing in soapy water and drying has not been changed. Until an entirely new concept of cleaning evolves, the current "wet-wash" home-laundry system will remain at about its present level of technical achievement.

MORE BECOMES BIGGER

In the areas of engineering design where MORE is the fundamental need, there is the consistent design pressure to "bigify" the design. Our engineering history is cluttered with the world's largest reciprocating steam engine, largest steam turbine, largest sailing vessel, largest steamship, largest dirigible, largest diesel engine, and largest jet engine. And now we have the world's largest jet airplane, the world's largest rocket engine, and the world's largest space satellite.

J. B. S. Haldane (6), years ago, wrote an interesting little essay, "On Being the Right Size," in which he notes that in nature all things evolve to a convenient size for the task to be performed in the environment in which they exist. All engineering designs have an optimum size within the constraints of materials, economy, and the basic concept. It has been characteristic for a design concept which has evolved through bigness to be made obsolete by the emergence of a new concept that fulfilled the need for MORE with a dramatic decrease in size. Huge steam locomotives gave way to smaller diesels of comparable power, 28-cylinder radial aircraft engines gave way to smaller, more powerful jet engines which gave way to smaller more powerful fan jet engines, etc.

THE ENGINEERING EXPLOSION

We are seeing the results of what could be called an engineering explosion. Our engineering know-how has expanded by orders of magnitude compared to what it was 50 years ago. Our scientific discoveries are increasing at a fantastic rate, and we have a diminishing "engineering gap." It used to take hundreds of years from the time a scientist discovered a basic principle to the time an engineer discovered a way to use the knowledge. In the early days of scientific activity, there were so few engineers and so few people involved in technological development that scientific discoveries were lost in the archives. They were reported by scientists and filed away. Some of these discoveries were rediscovered years later by someone who was unaware of the previous work. Technical ignorance made it impossible to capitalize on fundamental scientific discoveries. Limitations in materials, tools, and manufacturing techniques delayed the development of useful products from fundamental scientific discoveries.

But now, however, due to the availability of scientific knowledge to a sophisticated engineering community that is development-oriented, there is very little lag between a scientific discovery and its adaption to useful products. A typical example is the concept of the laser. Within five years of its discovery, a host of product ideas and applications using the principle were under development.

Every technical periodical reports on engineering developments involving the application of recent scientific discoveries which are under study. This is indicative of the type of engineering we're now doing and the new emphasis on conceptual engineering design and development. This creative engineering effort not only develops new products but accelerates the evolution of established design by the technological "breakthrough" that stimulates new markets by both technical obsolescence and reduced costs.

THE AGE OF DIME-A-DOZEN MIRACLES

This engineering explosion has created an age of miracles. Miracles are becoming so commonplace that they don't even rate any respect. People don't even laugh anymore when someone comes up with a wild idea. Within our own generation, products and services have been made available that weren't even conceived possible in the preceding generation, i.e., direct distance dialing, color television, stereophonic recordings, jet airline travel at supersonic speeds, etc.

Many of the seemingly fantastic proposals in science fiction literature have become commonplace in reality:

- astronauts floating freely in space suits hundreds of miles above the earth.
- satellites transmitting television pictures of the weather around the earth, and transmitting live televised events between continents.
- transport planes capable of 2,000 mph.
- submarines using atomic power that can circle the globe without surfacing.
- power plants developing electricity from atomic energy.
- rocket belts to propel individuals through the air.
- solar cells to convert sunlight into electrical power.
- laser ray guns.
- welding by light rays (lasers).
- rockets sending back pictures of the moon and the planets.
- electronic computers that control entire factories.
- robots working on assembly lines, etc., etc.

Engineering technology is at a place where the fantastic can be achieved tomorrow and the impossible only takes a little bit

"THEY"

How often have you heard the remark, "Why don't **they** come up with a ____________ so we can ____________?" Or, "We wouldn't have this problem if **they** hadn't built that ____________." Or, "Have you bought one of those new ____________ **they** have invented?"

Who is this "**they**" group? **They** are engineers—the community of designers, experimenters, and manufacturers who create and produce the products and services of our society. Engineers and engineering are fundamentally responsible for the physical conditions in which we live, our economic structure, our mode of thinking, our values, and even our philosophical and religious attitudes.

They are responsible for eliminating the economic need for slavery; for making education a necessity; for the virtual elimination of the struggle for survival. **They** have provided mass comforts, services, and leisure exceeding anything ever achieved by the wealthiest ruler in history. **They** have reduced the work week and have made leisure a mass luxury. **They** are responsible for releasing mankind from the survival struggle so that we are free to pursue perfection and explore our universe. If our society is ever to achieve the perfection of "the good life," it will be because **they** have provided the materialism that freed us from the struggle for survival so we can work toward it.

They, the engineers, are the people who make the scientific discoveries useful. To "know" is of no value unless there is someone who can convert "know" into "do."

longer. We're almost at the point where anything that we can conceive is achievable.

The engineering explosion has truly created an age of miracles. It has had a profound effect on superstition, resistance to change, and acceptance of innovation.

THE TECHNOLOGICAL IMPACT

The age of engineering and engineering design has had a telling effect on humanity. The Industrial Revolution and its accompanying technological evolution have made it possible for us to live in environments which would have been too hostile for survival before the advent of technology. It's difficult to assess which of the technological innovations has had the greatest impact. Certainly the technological advancements in transportation have contributed to our ability to sustain ourselves in hostile environments. It has made it possible to establish cities where food was not available, and to exchange goods and processes on a worldwide market. And of course, the availability of a wide market made it possible to expand production and make available goods and services to a larger number of people. The advent of refrigeration made it possible to preserve perishable foods and make them available to all parts of the country regardless of seasons. And, development of mass-production techniques using automatic machinery made it possible to produce large quantities of products at an economical rate thus making goods and services available to a mass market.

It is not important which of these innovations had the greatest influence. The important thing is that all of them have drastically altered human events and human society.

TOO MUCH OF A GOOD THING?

Have we already invented more than we can accommodate? What else do we need? Do we already have all the comfort we can endure? What else can be done for us that hasn't already been done? There is concern that the engineering explosion is creating more social problems than it's solving. Many of us are getting too comfortable, too well off, too soft. We may be getting to the point now where people don't need any more miracles. There are more than 25 fractional-horsepower electric motors in most homes. There are scores of mechanical devices in every home providing labor-saving convenience, comfort, and pleasure. For example, air-conditioned homes, cars, offices, and public buildings are available to nearly everyone. Telephone, television, stereo hi-fi, central heating, two cars, two bathrooms, garbage disposal, electric dishwashers, and automatic home laundries are common conveniences.

Automated processes are providing mass services at reduced cost and with reduced labor. The net result is a better living and a shorter work week. Engineers have provided society with the luxury of leisure. Recreation is a huge industry. Swimming, camping, boating, tourism, mountain and seaside resorts, and amusement parks are flourishing businesses.

How far can "the good life" go before we become swamped by the system? How much "good life" can we stand?

WHERE DO WE GO FROM HERE?

If some of us already have "the good life," what is left to be done? What new contributions can engineering technology make when we seem to have all we need? In spite of all we know how to do, it is unthinkable that everything has been invented. Inventions will cease only when human beings cease to be creative. And technological progress will continue as long as people can discover and learn.

The design horizons are broad and the problems to be solved seem to fall into three general categories:

1. Making the technological "good life" available to every human being.
2. Overcoming the massive problems being created by technology.
3. Continuing development of new scientific discoveries.

THE GOOD LIFE . . .

Engineering has created an almost fanatical worship of materialism. This country is living pretty high on the hog, and there are about 2½ billion other people in the world who want in on it. They are going to work to get it. They want to rise above the survival struggle, educate themselves, mechanize their society, and partake of "the good life."

The fantastic volume of products that will be required to meet the demands of the world's population will create staggering engineering problems, as well as economic and political problems.

An industrialized society and the accompanying standard of living consumes a prodigious quantity of raw material and energy. J. R. Weir (12) points out that the per-capita consumption of steel and oil in the United States is about 150 times that in India. To bring the rest of the world up to the living standard of the United

States would take annual rates of production of raw materials well over 100 times the present production. And the quantities of some of the more strategic metals (tin, copper, etc.) that would be needed would exceed all of the known ore reserves. The needed increase in power consumption would exhaust all the known fossil-fuel reserves within 50 years.

New materials, processes, and energy sources will most certainly have to be invented as substitutes for the dwindling natural resources. New techniques will have to be devised to extract minerals from low-grade ores or from the sea. Innovations in manufacturing will have to be devised to meet the demands of production and distribution. The product demand will create a monumental labor shortage and even more unique concepts in automation will be a necessity. Waste disposal, contamination, and pollution will become increasing problems as the world becomes mechanized.

A mechanized world that no longer relies on human labor as its power source will require huge increases in energy and mechanical power. The conventional power sources from water and fossil fuels will never sustain the demand. Engineers will have to devise new techniques to capitalize on solar energy rather than the fossil fuels. Engineering faces a tremendous challenge to figure out how we can all "live it up" and not use up all the "goodies."

YIPE! WHAT HAVE WE DONE?

A good-sized hunk of the engineering frontier is made up of the problems that have risen because of "the good life" that technology created. One of the most frightening problems, which is partially a result of "the good life," is the threat of overpopulation. The world's population is increasing at a geometric rate. Unless rigid birth-control measures are adopted throughout the world, the population could quite possibly outrun all of the physical resources that can be made available.

The consequences of uninhibited growth in population, in uses of energy and materials, and in appetites for technologically derived luxuries and comforts are upon us. All the seemingly endless supplies of conventional resources are suddenly becoming scarce and alarmingly finite—even terminal. The enormous pressures to maintain all the gains and all the promises of "more and better" are creating even larger demands on the need for more engineers and increasing engineering effort to deal with the consequences of past technological exploitation.

We are at a point in time that marks the end of an era of technological exploitation and affluence. We are entering an era of conservation and of development of long-term alternatives to an energy-intensive material-extensive society. Engineering know-how and effort will be focused on developing alternatives for energy sources, energy production, materials, reducing costs of manufacture, transportation, home heating, salvaging and reusing materials, reducing energy losses, increasing efficiency and productivity, reducing costs, etc.

> Engineer—
> Long on energy, short on cash
> Needs client—
> Long on cash, short on energy.

Engineering will be increasingly called upon to address the physical and psychological hazards that have developed due to the interaction of the mechanical world with the humanistic world. Engineers will be called upon to devise ways to protect humans from their machines. Noise, fumes, mechanical accidents, storage, repair, obsolescence, and disposal are increasing problems as the number and use of machines increase.

The engineer will inevitably become involved in solving humanistic problems such as mechanizing communication, information retrieval, education, and transportation. Scientific discoveries in the areas of biomedics have created a new frontier of bioengineering which will solve the problems of mechanizing vital organs of the human body.

The acceptance of engineering technology in our culture compounds the need for continuing engineering effort to sustain and improve our living standards. Our society has become accustomed to being served by the products of engineering design. There is a prevailing attitude, "If it's possible—let's do it." Society is demanding that the luxuries of scientific discovery be made necessities.

The price of technological affluence is paid in huge expenditures of natural resources and a deluge of trash, waste, junk, contaminates, noise, and congestion.

One of the great challenges to engineering will be the role of achieving a balance between the demands of society, the "fall-out," the conservation of resources, and the development of alternatives.

BEYOND THE BLUE HORIZON . . .

Engineering plays two fundamental roles; it adapts scientific discovery to solving our current problems, and, more exciting, it invents products and services that we didn't know we needed.

It is a new and imaginative type of engineering that is emerging. It could be called "conceptual design" or creative engineering. It is devoted to extending foresight into the realm of fantasy. A cult of highly imaginative, creative engineers who probe beyond the reaches of feasibility and peer into the technological future has emerged.

This facet of the design horizon is unlimited. The explorations of science into the heart of matter and the reaches of the universe are opening up unlimited and unimaginable opportunities for engineering innovation. The mechanization of our current space research vehicles and systems is a limitless field of engineering innovation and challenge. Similarly, the problems of mechanizing the exploration of our "inner space" pose equal challenges.

"The sky is the limit" in engineering design. The world will never have an excess of imaginative ideas. To provide solutions—for problems that haven't been discovered yet—is the brightest design horizon of all.

SELECTED REFERENCES

1. Alexander, T. "The Wild Birds Find a Corporate Roost." *Fortune*, Aug. 1964, p. 130.
2. Armytage, W. H. G. *A Social History of Engineering.* Faber and Faber: London, 1961.
3. Dalia, F. J. "The Engineer's Vital Role in American Economic Development." *Consulting Engineer,* May 1963, p. 102.
4. Finch, J. K. *Engineering and Western Civilization.* McGraw-Hill, 1951.

5. Gordon, T. S. "Some Crises That Will Determine the World of 1994." *The Futurist,* June 1974, pp. 115–121.
6. Haldane, J. B. S. *Possible Worlds.* Harper & Row, 1928.
7. *Here Comes Tomorrow–Living and Working in the Year 2000.* Dow Jones Books, 1967.
8. Narver, D. L. Jr. "What the Future Holds for Engineers." *Consulting Engineer,* Oct. 1964, pp. 12–122.
9. Newman, J. "1994: The World of Tomorrow." *U.S. News and World Report,* 1973.
10. Usher, A. P. *A History of Mechanical Inventions.* Harvard University Press, 1954.
11. Walker, E. A. "An Event Worth Celebrating." *Electrical Engineering,* Jan. 1962, pp. 27–31.
12. Weir, J. R. "Technical Manpower for the Next Hundred Years." *Electrical Engineering,* May 1961, pp. 338–340.
13. Whinnery, J. R. *The World of Engineering.* McGraw-Hill, 1965.
14. Wise, C. "Are Bigger Machines Better?" *Machine Design,* May 27, 1965, pp. 124–129.

2

Engineering Design . . . What Engineering Is About

The scope and variety of an engineer's work is extensive, ranging from sales and industrial management to rigorous research and experimentation toward the adaption of new discoveries to useful applications. In the middle of this spectrum is engineering design, the activity that most clearly defines what engineering is —the process of developing products and processes to satisfy a need, the creative problem-solving process.

> Are you disgusted?
> Your gizmo busted?
> Well, help is here!
> An Engineer!

RESEARCH–ENGINEERING–MARKETOLOGY

Engineering design depends heavily on three areas of knowledge. These are scientific information, engineering technology, and the body of knowledge that could be called "marketology."

Engineering design activity is an imaginative integration of each of these areas toward the development of a useful product.

Scientific information defines what we can do—what nature will allow. This is the body of facts, theorems, laws, and information that has been and is being compiled from basic research activity. It is this basic information, undefined in its usefulness, that is available for engineering application.

A great proportion of our current scientific information is being discovered through the efforts of an activity called *engineering research.* This activity is the work of scientists with engineering orientation, who explore nature to obtain answers to specific problems created by a design need. It is part of the developmental activity surrounding the evolution of a product. Many products have attained the highest degree of sophistication that present knowledge will allow. Further progress sometimes must wait until shortcomings in knowledge are overcome through research.

The Engineering Design Spectrum

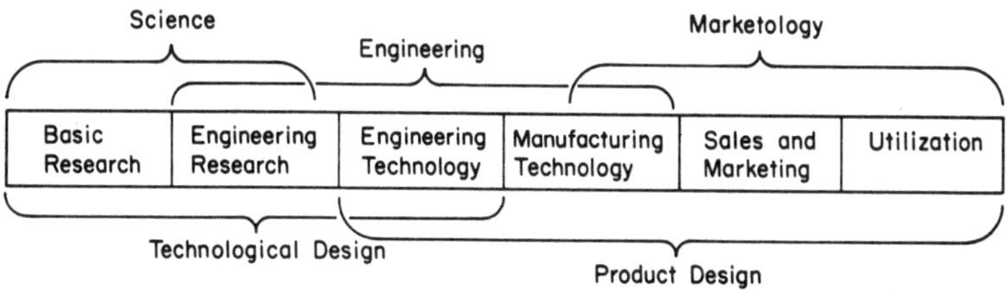

Engineering research is becoming a vital facet of the design process. The expansion of this activity throughout industry, along with the demands of space technology, is creating a great demand for engineering students with advanced degrees.

Engineering technology is the how-to-do-it phase of the design process. It spans the gap between the literature and the shop and encompasses a vast body of specialized techniques. This is the field of synthesis, analysis, experimental data, formulas, charts, diagrams, computers, etc., which converts knowledge into working hardware.

Marketology is the study that relates an engineering design concept to the practicalities of mass production and consumer acceptance. It is a technology of production, sales, distribution, and finance which surrounds the manufacture and marketing of products. This field, spawned by the pressure of competition, has become an important facet of the engineering process. It has brought the influence of sociological, psychological, and economic ramifications to bear on design decisions. It relates consumer acceptance, product usefulness, and costs to the design and manufacturing decisions. The automobile industry is typical of the role marketology plays in the design process.

Obviously, the engineering design process is a team effort by many specialists. Design engineers have to call upon, and weigh, information from many sources to bring about a successful design. To be naive about the broad ramifications of the design process would be unforgivable in the modern competitive field of engineering.

The entire scope of an engineer's design activity can be divided into two basic kinds of design effort: creating a unique technically-oriented solution to a specific problem, which could be labeled *technological design,* and developing a product that will be useful to a large number of people, which could be labeled *product design.* In technological design, the engineer is essentially a consultant to a client who has a problem. The vast majority of all engineers work in this capacity. In product design, the engineer is

MARKETOLOGY
The Science of Mass Production

The design of products that are mass produced for a large consumer market is greatly influenced by the number of units to be made and the reaction of the consumer. Marketology is the business of making design-influencing judgments concerning:

- consumer needs vs. wants vs. fads vs. whims.
- effect of the product on community life.
- effect of the product on international trade.
- effect of the product on labor and job security.
- consumer safety and health hazards.
- availability of materials, supplies, processes, and production capacity.
- consumer resistance and/or reaction.
- distribution—sales—service.
- product life vs. replacement vs. repair.
- appearance: effect on consumer; effect on community.
- scrap, congestion, pollution, storage.

a specialist among a team of specialists in styling, marketing, manufacturing, etc. The engineer's role is to provide the technical alternatives to meet a wide variety of objectives, with cost and profitability being the overriding goals.

> You're saying the thing's overheating?
> It's costing a bundle to run?
> And the Widgit keeps jamming the Whatzit?
> And no one knows what's to be done?
>
> My Friend, what you need's a solution
> Devised by a good Engineer
> Who'll whomp up a clever contraption
> And wipe out your worry and fear.

TECHNOLOGICAL DESIGN—CLIENT-ORIENTED DESIGN

There is a vast amount of engineering design work that the public never sees. It includes the machinery, processes, and specialized apparatus used in industry and the "one-shot" developments that solve particular technological problems in transportation, defense, research, exploration, etc. The automation of manufacturing has created thousands of singular innovations in specialized machines designed for particular needs in particular plants. Packaging machines, process machinery, machine tools, etc., are all custom designed and built in extremely limited quantities (often only one). All of the complex hardware for space exploration, undersea exploration, and military ordnance fall into the scope of technological design.

Actually, this area of design engineering is an engineer's utopia—it involves creating and inventing unique solutions for specific problems. Function is the overriding criterion; cost is secondary. Function is never compromised for cost—cost is justified by function.

Technological design depends heavily on research for breakthroughs that solve its problems. It is design activity that is carried on by the Research and Development (R & D) groups in industry. A design emerges through a custom operation—designed by a team, constructed by a group of skilled craftsmen, tested and debugged, and installed. It is a tailor-made project, not unlike the design and construction of a building or bridge.

This field has a great deal of glamor because it creates fantastic machines today and impossible ones tomorrow. Technological design produces equipment to drill a hole through the earth's crust

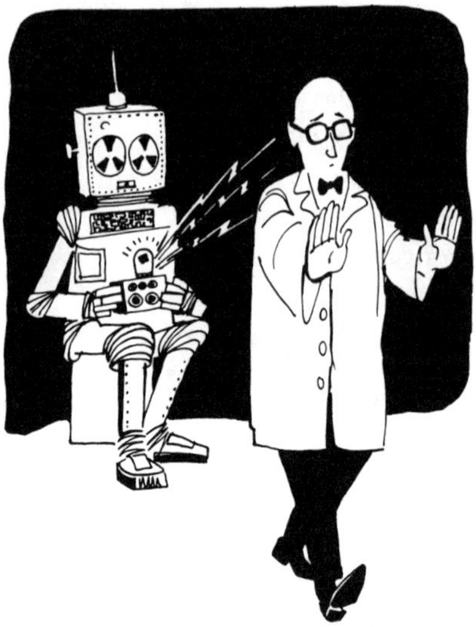

. . . it is the activity that creates the fantastic machines today and the impossible ones tomorrow.

under several thousand feet of water, to bend pretzels, to machine an engine block automatically, and to automatically assemble and weld the body shell of an automobile. Engineers in this design activity have invented a strip-mining shovel to scoop 90 cubic yards of material at one bite and a space probe to send back photographs of Mars. Technological design combines conceptual engineering and engineering research. This field produces the innovations in engineering technology; it is the breeding ground of technological progress, product evolution, and advanced techniques in design and manufacture.

PRODUCT DESIGN–CUSTOMER-ORIENTED DESIGN

Product engineering is designing for mass production–making a retail product for a mass market. It is a customer-oriented design activity in which marketology has an overriding influence on design decisions and in which cost is the principal criterion. The creative effort in this activity is devoted to the massive problems of creating a competitive product with superior market appeal, at the lowest possible cost, in a large number of units.

The success of a product depends heavily on whether it can be produced in volume at reasonable cost. Often the success of a product depends completely on the development of specialized manufacturing techniques. The product designer frequently is forced to devise unusual or unique schemes for assembly, fastening, molding, etc., which will achieve product function and service-life, yet simplify the process of manufacture and hence reduce cost.

While technological design is oriented to getting a product to do the job, the product design engineer not only must design a functional product but must create a product that can be manufactured in great quantity. Volume production creates many problems of considerable complexity and magnitude. For example, consider the effects of cost when 100,000 units are involved. Every penny saved or lost, per unit, amounts to $1,000 saved or lost on the entire production. Even scrap becomes important. Every drilled hole not only costs money but converts the metal in the hole from steel worth 20¢ a pound to steel scrap worth 2¢ a pound. If every unit produced one pound of scrap, the entire production of 100,000 units would lose $18,000 just converting a portion of the raw material to scrap.

Since product design is a profit-oriented mass-market activity, marketology plays a large role in influencing design decisions. Design criteria evolve from concern for appearance, safety, usefulness, repairability, product life, and consumer acceptance. These considerations must be balanced by costs. But costs are primarily influenced by the method of manufacture and the number of units to be made.

When you think about it a bit, the major engineering effort in product design is primarily concerned with how to make it and how to do it at low cost. However, the functional usefulness of the product cannot be ignored. A great amount of creative effort is necessary to maintain a competitive position. It is a continuous process of technical evolution through modification and technical innovation. The mass-market product will fail unless it is continuously updated to incorporate new materials, new trends in consumer needs and preferences, and new functional concepts. Sensitivity to technological design progress is vitally necessary for survival.

PURPOSE OF DESIGN
A Technological Solution to a Need:
- To create a **physically realizable** object
- To create an object with **economic worth** and **financial feasibility**
- To choose an **optimum alternative** solution
- To create a useful product
- To develop a new way
- To reduce cost, inconvenience, hazard
- To produce convenience, service, value

As important as technical innovation is to product survival, the product designer is faced with the ironic necessity of overcoming consumer resistance to change. This necessity and the strangling constraints of cost are the primary causes of the gap between technological design innovations and mass-produced consumer products.

THE TECHNICAL GAP–KNOW-HOW VERSUS USE

There is a substantial (although diminishing) time-lag between breakthroughs and design innovations that emerge from Research and Development projects and their subsequent adoption in mass-produced consumer products. We frequently read of experimental automobiles with exciting and useful new features. However, it takes years for many of these innovations to be made available to the mass market; indeed, many never become available.

Over 30 years ago, a proposal was made to install a polarizing lens on automobile headlights and polarizing grids in the windshield to reduce the hazards of headlight glare. Recently, solenoid sensor switches have been developed to dim headlights automatically when meeting oncoming cars. Neither of these ideas has become a standard safety feature on automobiles. Color television was developed and made available some 10 years before it won sufficient industry and public acceptance to cause people to "trade up" from black and white. The principle of electric automobiles is well known, and many versions of electric-drive vehicles are available. However, there still is no movement toward mass marketing of electric automobiles. Eventually, they may replace combustion-engine-driven vehicles, but there are many hurdles to be overcome.

There are a number of realistic and legitimate reasons for the delay or prevention of innovations in mass-produced products:

- **Production limitations:** New products are delayed because the mass-production techniques required to justify production economically have not been devised. The conversion of home stereo-music systems from mechanical reproduction to magnetic tape has not yet been accomplished because a technique for mass producing recorded tapes has not been perfected. The sculptured styling of automobiles could not be accomplished until a deep-drawing body steel was developed. Countless product innovations have failed to reach the market place because manufacturing technology has not been available to produce them. (I venture to speculate that there is a fortune to be made from old patent ideas. Many ideas, at the time of filing, undoubtedly were ahead of their time and couldn't be mass produced economically. Some of these ideas might now be economically feasible within our current technology. There's a bonanza waiting there!)
- **Present commitments:** Many innovations cannot be justified because they would require a massive overhaul of an entire production system; retraining of hundreds or thousands of technicians, service people, or operators; scrapping of huge parts inventories or backlogs of current production; etc.
- **Sociological implications:** A proposed product innovation may require drastic overhaul of safety codes, regulatory laws, or insurance rates. Or, it may cause real or imagined hardships on the labor market, create health hazards, upset foreign trade agreements, or create economic and competitive pressures.
- **Consumer resistance:** The public has to be educated to want and use a new product idea. Market surveys may show there is not enough consumer interest to justify the production techniques required to lower the cost of manufacture to an acceptable level. The consumer is slow to change from his

present way of doing things. He has to be "sold" by a massive promotional campaign to demonstrate the new product's value.

- **Lack of capital**: Many companies cannot justify the gamble of tooling up and manufacturing the product without a profit for the time required to win customer acceptance. The availability of mass-market color television is largely attributed to the single-handed gamble of one company and its affiliated broadcasting network. They made the only sets and produced the only color programs for a number of years, at great financial risk, until public acceptance encouraged the entire industry to promote color television so that it could be made available to everyone.
- **Lack of competition**: In some cases, an industry has won an exclusive patent—in effect, a market monopoly for a product innovation. Since there was no competition, there was no motive for improvement of the product—and useful innovations were never made available to the public. The Model-T Ford had a virtual monopoly of the utility-automobile market; for many years, the car remained unchanged (and available only in one color—black). The formation and growth of General Motors has been attributed to a policy of creating automobiles with up-dating innovations that ended the Model-T era.

The engineering problems involved in the continuing evolution of mass-produced products are complex and range over a broad spectrum. They do, however, illustrate the scope of concern that an engineer must face in the design process. It is not enough to limit engineering design to technical problems. The impact of a mass-produced product on the entire social and economic structure of the nation continually affects the design decisions that must be made.

WHERE DOES IT ALL START? (WHAT'S THE PROBLEM?)

The engineering design process cannot begin unless there is a need—a question to be answered, a problem, a difficulty, a motive. Since design is a problem-solving process, there is a necessary and important initial phase of seeking and choosing the problem.

There is a generous portion of design activity that evolves from self-generating problems. That is, the normal evolution of

There is a generous portion of design activity that evolves from self-generating problems.

the use and requirements for the product presents a continuing sequence of new problems to solve. However, ever-expanding technology and demand for "the good life" have spawned a problem-seeking urgency within the engineering design field.

Many industrial concerns now have organized "brainstorming" sessions within their engineering organizations for the express purpose of defining new product needs and problems. Consulting firms have been created to search for product needs for companies wishing to expand and diversify in the face of declining markets for outdated products. A few large companies have hired extremely creative people specifically for "crystal-ball" prognostications.

There are several schools that have organized a curriculum in product design with particular emphasis on searching for new product needs and ideas. Marketing consultants make it their business to advise industry of the trends, needs, and requirements for new products. Many companies now maintain magnificent research facilities staffed with highly skilled scientists and researchers who, in addition to their basic research activity, explore the frontiers of product possibilities and consumer needs.

Farsightedness and clairvoyance are demanded of the engineer. Besides being able to solve the problem, the engineer is being called upon to invent the problem in the first place. There is a lucrative career for the engineer with visionary imagination.

IS IT POSSIBLE?

When a design need is expressed, it is usually a vague sketch of a proposition. Most often, it is expressed as a fancied idea or an expression of a dream—"Wouldn't it be terrific if we could just push a button and . . .?" A product need must be explored, rephrased, and cast into terms that define the basic problems which must be solved.

THE HEADWATERS

Where do design projects come from? Creative engineers don't have a corner on the idea market when it comes to "why don't they . . . ?" ideas for the development of new products. Here are some typical sources of design project ideas:

- Newspapers, journals, and magazine articles describing new developments
- Purchasing agents
- Backyard inventors looking for industrial support
- Venture capital organizations and commercial bankers
- Patent attorneys representing clients
- Consulting engineering firms
- Sales organizations
- Marketing consultants
- Research organizations
- Trade associations
- Community industrial development commissions
- Patent office records
- Advertising agencies

A feasibility study must be made first. This involves much fact gathering and survey work to bring to light all the conditions, motives, and constraints that interact to create the environment within which a design must be created. Then, a creative ideation search must be made for ways to accomplish the original proposition. This search is followed by an analysis to determine what conditions are necessary to tell if a design idea could even exist within the limits of natural laws.

For example, suppose a television station desires a way to extend its coverage over a 1,000-mile radius from a single transmitter. A feasibility study must be launched to determine how such a need could be fulfilled. Several ideas evolve. Suppose one is the proposition to build the world's tallest antenna tower—perhaps over a mile high! This idea must be studied to determine

THE DESIGN PROCESS
A Sequence of Dilemmas

Advanced Planning:
- What changes can we expect and what will they do to our products?
- What do consumers need that they don't get now?
- What's new that we can adapt to a new product?
- What problems still need solving?
- How can we create an entirely new product?

Feasibility Studies:
- Will the idea work? Can it be done?
- What will current technology allow us to do?
- Can we possibly make it?
- Could we make it practical?
- Could we sell it and make money?

Preliminary Design:
- What's the best way to do it?
- How can we make it reliable? Stable? Strong enough?
- Will it stand up under all conditions of use?
- Can we optimize its performance?
- Can we simplify it—cut costs?
- Is it compatible with the environment?
- What materials must we use?

Production:
- What shape must each part be?
- How will we manufacture each part?
- How will we assemble it?
- What about final appearance? Style? Finishes? Colors?
- Do we need to devise special machines and tools to make it?

Marketing:
- Should we provide model variations?
- How about repair, maintenance, and service?
- How will we package, ship, and distribute the product?
- What about installation? Do we need to provide manuals? Train installers? Provide special equipment for testing, adjusting, and check-out?

if it is even possible to build a structure of such height, and then studied to estimate its practicality. In addition, the idea must be examined in an effort to predict the design problems that might evolve.

The feasibility study is a necessary step to gain sufficient confidence in an idea to warrant the time, money, and effort required to make a detailed design. Intuition often is not enough to test the value of an idea. What may appear to be a totally ridiculous idea, one that couldn't possibly be achieved, may well turn out to evolve into a revolutionary approach.

Considerable creative effort is required in the feasibility study, to ensure that the idea be treated fairly and without bias. During the evaluation of the idea, it must be manipulated to assure that all alternative considerations for achieving the requirements are accounted for. The criteria upon which a judgment is based must be evaluated to see if they apply to the new situation.

We asked a group of students to determine if it is feasible to use a flywheel (with an upper limit on its diameter) to be an energy source for a given power requirement. Some students checked the idea by using the recommended code limitations on flywheel speeds and rim stresses and immediately concluded that it was impossible. Others, however, concluded that high-strength steels and a special design configuration could easily exceed the power requirement. The students using the code were defeated because they arbitrarily assumed that the flywheel would have to be made of cast iron.

A feasibility study is an iterative process—check and revise, then check again—until all possible alternative ideas for achieving the requirements of the design are explored. This is the idea-getting and idea-evaluating stage of the design process. The ultimate goal is a product idea that has enough merit to be worth extensive design, development, and testing. However, it is also likely that a feasibility study will verify that the idea would not be practical in terms of cost or utility; that is, "Sure we can do it—but is it worth the effort?"

LET'S GO AHEAD

If the feasibility study indicates a promising design concept, with all the attributes that are desired, a decision is made to go ahead. Now a preliminary design is developed and extensive surveys are made to define all the requirements. A configuration is devised and analyzed. Parts are sized for strength and function;

computations are made to predict the variation of all important parameters; and a design for a working model is evolved.

This is the development stage—creating the preliminary design that leads to the construction of a working model and to actual performance tests. The process of selecting the best alternatives still continues—but the decisions are concerned with components and sub-systems within the total design. The complexity of the process increases, and the original concept now becomes, in fact, a multitude of individual design problems. Each is dependent upon the function of the others, and one decision affects many other preceding and subsequent decisions. The process of development generates a complex series of interacting design decisions.

The motive during the preliminary design phase is to establish and verify that a practical working design can be achieved—that it can be produced at a reasonable cost—that the product idea is acceptable for mass production and the market. This phase, in addition to extensive design activity, usually includes a program of prototype construction and performance testing. Depending on the product and its market potential, the testing program may range from simulated-use tests in the laboratory to extensive field tests, including actual performance trials by potential customers.

The prototype stage is an important decision-making point in the design process. At this time, not only must functional capability be proven, but the product's commercial image must be demonstrated. Some of the basic concerns that "bug" the design team at this point are:

- Will the customer be satisfied?
- Will costs create a selling price that will produce sufficient volume?
- How about its appearance?
- Have we thought of everything?
- Will it stand up well under actual operating conditions?
- Is there any possibility of poor service life?
- Can the customer operate it without difficulty?

Most of these questions are answered by building one or more final prototype models of the product. These final prototypes are exact this-is-it copies of the product as it will be sold to the cus-

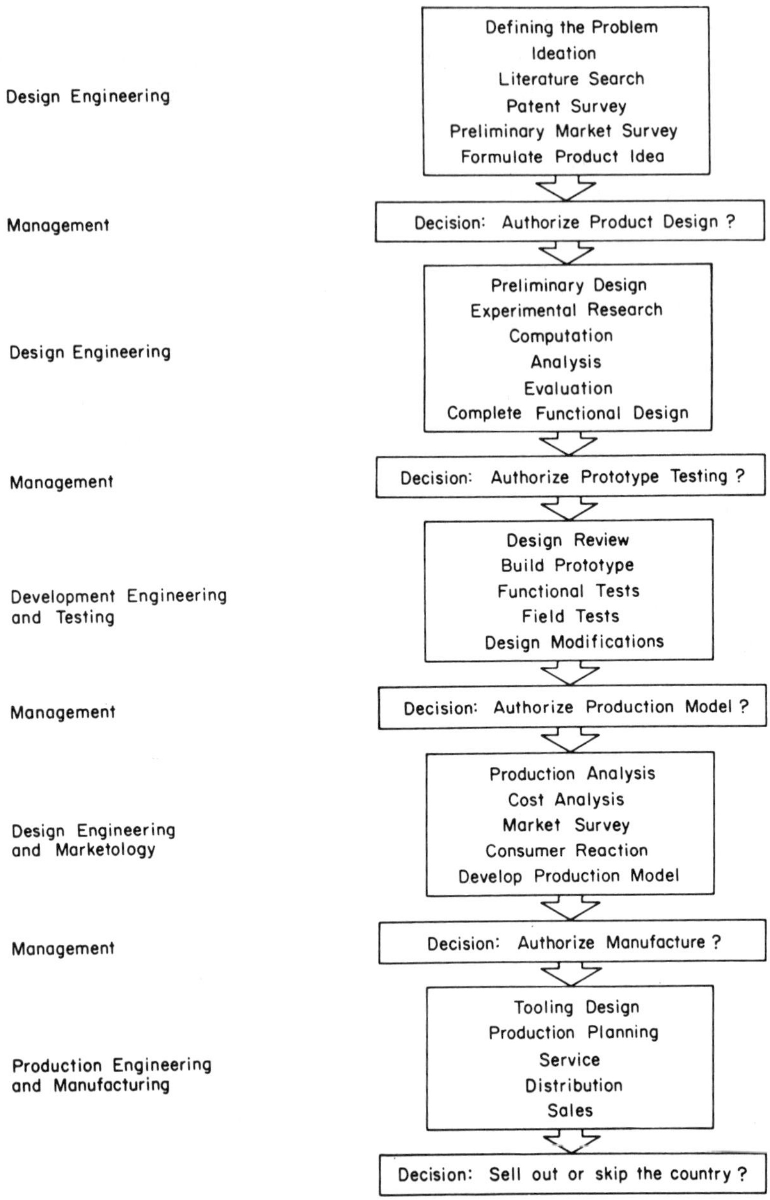

PRODUCT DEVELOPMENT

Decisions! Decisions! Decisions!

tomer. Extensive actual-size tests are performed by inviting potential customers to use the prototype and studying their performance and reactions. Studies are made to determine the reaction and comments of sales people, dealers, distributors, production people, and top management.

All through the development stage there is a continuous process of modification and refinement. The prototype stage is extremely fruitful in identifying the refinements and changes that assure customer satisfaction. Each proposed change and modification must be weighed for its potential effect on the methods of manufacture and the product cost.

The decision to turn the product over to the production people is a difficult top-management judgment. By now, the design people can justify its function and customer acceptance, but the problems of manufacturing and the ultimate cost have not yet been determined. The decision to go into production weighs heavily on the expected market. The market determines the volume, dictates the investment in tooling, the scope of the tooling, and drastically affects the ultimate cost. Judging the potential market is, at best, a dilemma that requires an intelligent guess based on the judgments of many experienced people in the company. As one top-management man put it, "As long as we're right 51 percent of the time, we're in business."

OKAY, LET'S TOOL UP

The preparation of a product design for volume manufacture requires a considerable amount of engineering design effort in great detail. Every piece and part of the design must now be reevaluated from the standpoint of how to make it. Complex problems evolve regarding the cost comparisons of different methods of manufacture as they relate to the volume required, the time needed to make the product, the complexity of shape, the type of material, the precision of dimension, and surface finishes. Explorations have to be made to determine if several parts can be combined into one single part or subassembly. Problems of fastening and assembling parts arise, and decisions regarding the development of automatic assembly techniques vs. hand operations have to be considered. Concern is given to the relative costs of (1) making the parts (2) buying off-the-shelf components, or (3) subcontracting components and subassemblies to specialized manufacturers.

Invariably, explorations into the problems of manufacturing produce advantageous alternatives that can be achieved only by a modification of shape, material, appearance, assembly and/or function of the original design concept. Such conflicts require reconsideration of the design by the development engineers. It is quite likely these conflicts will present formidable problems and will demand many compromises. The demand for design modifications never ceases throughout the entire design process. Each new consideration creates a new conflict which must be resolved.

THE DECISION MAKER'S SNAKE PIT

What monsters lie in wait to threaten the decisions to be made during the design process? Here are some of the circumstances that could wipe out your new product:

- Competition will initiate quick improvements to weaken your needed share of market.
- A research "breakthrough" will produce an advanced design that will capture all the market.
- New technological developments will eliminate the market for this type of product.
- The pace of technical developments will make the product obsolete before you can sell enough to write off costs.
- The time it takes to get into production will allow a competitor to beat you to the store.
- A foreign competitor will undercut your price.
- A design defect will not become apparent until the product is in wide distribution.
- Customer preference will change.
- The market survey was not accurate.
- The cost of your manufacturing operation will rise and wipe out your competitive position.

The production phase of product design can easily generate many formidable design problems when the production volume necessitates a high rate of production. Frequently, special machines must be designed to automatically produce the parts or assemblies. Extensive production planning studies must be conducted to develop the sequence of production, the flow of materials, the logistics of supply, storage, and transportation. Jigs, fixtures, special hand tools, material-handling equipment, processing machinery, inspection facilities, and final test procedures and equipment must be planned for and installed. Finally, to expedite the product out the door, packaging, labeling, and shipping procedures must be evolved and special handling equipment must be designed.

It is during this phase of design planning that cost and time become the overriding criteria. Scheduling all activities to achieve a specific production date involves establishing lead times for all phases of manufacturing. The design engineers responsible for the design and construction of specialized machines are required to work to rigid time schedules and due dates. No other teams in the entire design process have to work under such severe constraints of time and cost as do the production design engineers. The initial design phases were constrained primarily by functional requirements. Production design is concerned primarily with how to make it at the lowest possible cost, and as soon as possible!

IN OTHER WORDS . . .

If one had to boil it all down, engineering design is a trial-and-error sequence of choices among a number of alternatives, in which each decision is affected by compromise between a number of conditions and constraints. It demands (1) meticulous attention to detail, (2) the coordination of a wealth of information, (3) the search for a variety of ideas at every stage, and (4) an over-all necessity to achieve the best performance at the lowest cost in the shortest possible time.

It is a team effort enlisting the talents and resources of scientists, engineers, technicians, machinists, manufacturers, administrators, lawyers, accountants, salespeople, merchants, and public relations experts. But above all, it is a problem-solving process that creates something that is needed and useful.

SELECTED REFERENCES

1. Alger, J. R. M., & Hays, C. V. *Creative Synthesis in Design.* Prentice-Hall, 1964.
2. Gasson, P. *Theory of Design.* Harper & Row, 1973.
3. Gibson, J. E. *Introduction to Engineering Design.* Holt, Rinehart & Winston, 1968.

4. Glegg, G. L. *The Design of Design.* Cambridge University Press, 1971.
5. Glegg, G. L. *The Science of Design.* Cambridge University Press, 1973.
6. Hill, P. H. *The Science of Engineering Design.* Holt, Rinehart & Winston, 1968.
7. Knick, E. V. *An Introduction to Engineering and Engineering Design.* Wiley, 1969.
8. Love, S. F. "The Mastery of Engineering Design." A twelve-part series, *Design Engineering,* Feb. 1977–Feb. 1978.
9. Love, S. F. *Planning and Creating Successful Engineering Designs.* Van Nostrand Reinhold, 1979.
10. Marvin, P. R. *Planning New Products.* Penton, 1958.
11. McCune, F. K. "What Makes a Good Design Good?" SAE paper, 1963.

3

The Design Process . . . Doing Engineering Problem Solving

The crux of the design process is creating a satisfactory solution to a need. The need may be to improve an existing situation or to eliminate a problem. Or, the need may be to develop a use for a new discovery or concept. In any case, it is what engineering is all about—using knowledge and know-how to achieve a desired outcome. Designing is problem solving. It is *creative* problem solving. A change has to be created. Something different will be produced, and someone will have to decide what to do with the result.

> Love Engineers—
> They've always got problems

Engineers are applications specialists. They apply the principles, discoveries, experiences, techniques, and methods derived from ages of research, experimentation, trial and error, and invention. It really takes some doing to determine which elements of this vast storehouse of knowledge apply to the situation at hand. Developing the best combination of principles and procedures into a highly desirable plan and/or product is an engineer's business and it's tough work.

The design process is not just a matter of dreaming up clever ideas and schemes. It is a long process of imaginative planning, detailed analysis, computational prognostication, experimentation, detailed sizing, specifications for every piece and part, development of tools and manufacturing procedures, selection of materials, and detailed planning for assembly, maintenance, repair, safety, durability, sales appeal, etc. And, superimposed on all of this activity is the continuous and necessary attention to cost. The design process is an integration of technical knowledge drawn from the research laboratory and applied to the market place and customer use. It converts information into decisions and ideas into useful hardware.

LET'S GET RIGHT DOWN TO IT . . .

We can devote a lot of pages to discussing all the kinds of things that are involved in the design process in engineering. After all, it encompasses everything an engineer is apt to do. It is, in fact, the engineer's total territory and job description. But what is really important is a discussion of how an engineer does and should go about designing, problem solving, and solution creating.

Brain for Hire!
By the job
Or by the hour
Thoughtmanship Guaranteed

HAVE BRAIN–WILL TRAVEL . . .

Let's face it—an engineer is a brain for hire. An engineer working for a salary or a consulting fee is under hire to a *client* as a problem-solving specialist. Only engineers who are independently wealthy (or are college professors) have the luxury of designing something just for the fun of it. Since engineering design is essentially a client/consultant arrangement, there are some fundamental rules of the game that are at work—always—anywhere.

Rule 1. *Listen to the client—then find out what really needs to* *be done.* Clients have problems and needs—their descriptions of what is wrong and what they want may be a million miles away from what the design engineer ends up having to do about it.

Rule 2. *Maintain a project log.* Log in everything that was said, read, collected, decided, created, or rejected . . . in ink . . . dated and signed. The solutions engineers create inevitably involve a great deal of capital expenditure, income, human safety, and the professional prestige and job security of a lot of people. A well-kept record of all the thinking and deciding that went on is mandatory to sort out ownership, responsibility, and blame or fame.

Rule 3. *It's the clients' problem—it's their nickel.* The clients have to make the decisions. It's the engineer's job to advise them of the alternatives available and their relative merits. Therefore, the burden is on the design engineer to provide the feasible relevant options and a complete evaluation of all their characteristics.

It's not nice
To fool Mother Client!

Rule 4. *They've got to believe you.* If your clients get the slightest idea that you don't know what you're doing, you and everything you've done are through—turn in your badge and go home. It is imperative that you research and document your sources. It is imperative that your recommendations are backed up with facts, experience, logic, sound analysis, and rationale. They asked you to do it because they don't know how or don't have time to figure it out themselves. If they don't believe you, they have to turn it over to someone else until they can get a solution they can believe in and risk spending money to carry out.

Rule 5. *You have to sell it—it won't sell itself.* You may have found the greatest solution since sliced bread, but if your client (or her or his decision maker) can't recognize it, it is totally useless. The chances are good that most of the design engineer's clients are not engineers themselves. Or, if they are engineers, they probably are not specialists or up-to-date in the technology of your solution. To plunk down a highly technical engineering report will get you nowhere. It is imperative that you prepare a report of your design/solution in terms the client can understand and *wants to hear.*

O.K.! HOW DO DESIGN ENGINEERS DESIGN?

With considerable difficulty! They don't follow any precisely defined and rehearsed procedure. Most muddle along getting somewhere one day, losing ground the next, giving up, starting over, hanging in there until it finally works out to something that offers at least a meager sense of accomplishment. It's a laborious process involving a lot of redoing, changing, trials, and disappointments.

Idealism is incessantly compromised. The one common attribute of all engineering designer/problem solvers is dogged tenacity and an unshakable belief that *there is* a way.

Although all design engineers have their own styles, they all end up doing the following (one way or another).

Defining the problem. Doing the detective work to find the cause of the effect. The clients' definition of the problem is essentially a description of what they want and are not getting. This phase involves interviewing everyone in the outfit associated with the situation, studying existing operation data, collecting new data, visiting the site, and observing what happens before, during, and after. The objective here is to find the critical parameters that must be controlled or changed to achieve the desired outcomes. Once these are known, a search must be made to learn as much as possible of what is already known about such matters—the Inevitable Information Retrieval Phase.

> A design engineer's a detective
> Who's searching all over the town
> For clues to the client's dilemma
> Of why the contraption broke down.

Information retrieval. It will be a rare occasion that the design engineer will be the world-renowned "know-it-all" expert in the technology needed to solve the client's problem. It will be a

The Nine Stages of Design

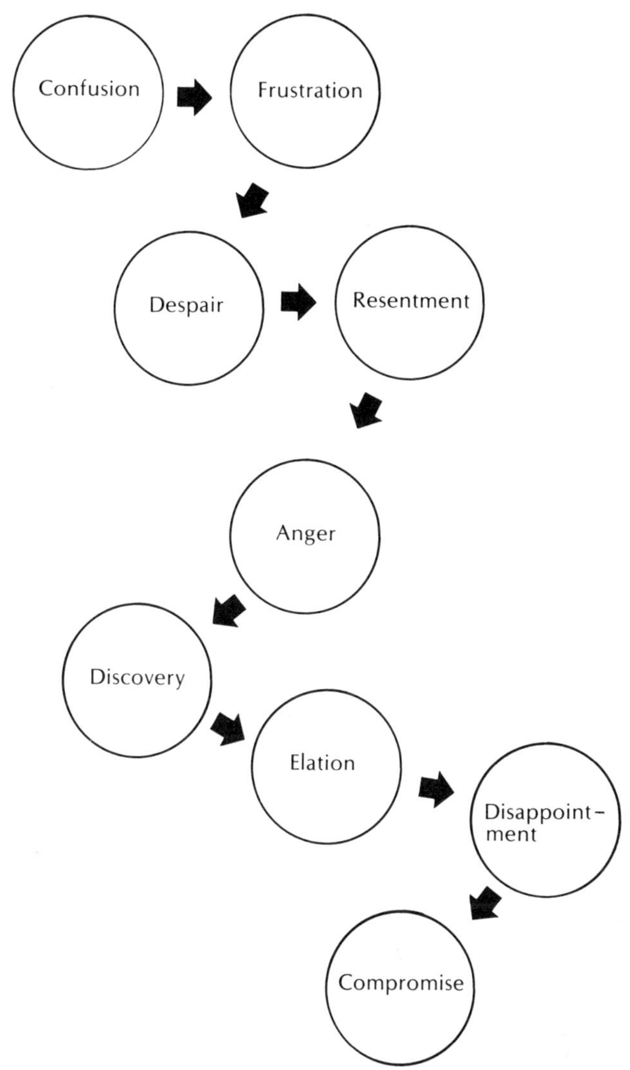

last resort "after all else has failed" . . . "we've GOT to get a solution" circumstance that will compel any client to seek out and pay for the services of the world-renowned specialist. It is almost a certainty that the design problem of any client will be dumped on the most available engineer within the company or within the neighborhood. All this is to say that nearly every engineer who gets a problem will NOT be expert and stands a good chance of not even knowing a thing about the specialized technology and knowledge it will take to solve it.

Therefore, one of the inevitable steps in the design/problem solving process is "getting boned up" on the available knowledge and experience relevant to the problem area. Any design engineer who expects to obtain a successful solution will do some or all of the following:

- conduct an extensive literature search—especially within the last five years' publications—and make use of one of the computerized bibliographic services;
- contact some of the authors of pertinent recent publications and seek their opinion and counsel about specific issues;
- seek out engineering research specialists in industry and academia to confer with them about the problem situation; and
- contact companies who specialize in the manufacturing of products applicable to the problem situation to learn the state of the art.

Seeking alternatives: The creative process. There comes a time in the problem solving process when sufficient knowledge and understanding of the situation have been obtained and one must start searching for, and compiling, a healthy list of applicable alternatives. This is the Creative Ideation Phase—the process of getting ideas for alternative ways to accomplish the whole solution and to construct the apparatus to pull it off. This is such an important and critical phase of the design project that we have devoted two chapters to the process (see Chapters 4 and 5).

The Creative Problem Solving Process is crucial because it is at this point that decisions must be made on the ultimate solution strategy. The outcome of this decision is a commitment to doing a lot of detail engineering design work to establish the worth and credibility of the chosen idea. A poor decision here means a lousy outcome later and a return to square one for another idea, a different approach.

The design engineer has three alternatives in the search for THE WAY:

- generate a unique, custom-made solution "from scratch,"
- find an "off-the-shelf" solution that someone else has used to solve a similar problem, or
- find a bunch of existing stuff, combine it, and adapt it.

A good design engineer will explore all three and show the clients what they would get and what it would cost to go with each of these three.

Development/synthesis/analysis/detail. Here is where engineers put to use all their basic engineering know-how. This is really the Design Feasibility Phase of the project. The outcome of the creative, alternative-seeking phase was essentially an intuitive decision that alternative "X" *ought* to be the best way to go. The next obvious step, then, is to do all the calculations—the analytical and experimental things necessary to show that alternative "X" can or cannot be achieved. If it can be done (i.e., in an engineering sense, nature will allow it and materials and manufacturing techniques are available to build it), then the designer has to conduct a cost/benefit analysis to see if it is worth doing.

There is much to say and discuss about this phase of the project. There are many fine books available on engineering design that are essentially devoted to all the techniques and procedures that can be used to establish the engineering feasibility of a design idea. Better than three-fourths of an engineering curriculum is devoted to the methodologies of synthesis and analysis of engineering systems. Most engineers are adequately equipped to do this phase of design. Actually, many engineers are quite content to do *only* this detail analysis and synthesis phase of a project and let someone else worry about all the rest of the design process.

There is, however, a psychological dimension to this phase of the project that tends to be an "alligator in the moat." There is a heavy engineering work load involved here. Much time and effort is invested in computation, computer modeling, and analysis procedures. It is inevitable that a tremendous resistance begins to develop against any notion to abandon it when the data begin to show the idea is not good enough or even "not worth a toot." There is a strong urge to salvage "all that work we did" by patching, modifying, forcing, adding on, etc. When the design engineer begins to "force-fit," the inevitable result is complexity, compromised objectives, reduced safety factors, increased costs, and all sorts of other bad consequences.

A not-so-good result at this stage is a clear signal to go back to one of the other *different* approaches and run a design analysis of it for comparison. As a matter of fact, the most successful design studies are those that have included a design analysis of several different alternative approaches. The design engineer can then present the clients with several well-defined options and show them the explicit consequences of choosing among them. Remember: it's the clients' nickel. They have to decide and you have to advise them of *all* the feasible alternatives, including the poorer ones.

Cost/benefit analysis. Most engineers would be quite content to be "done with it" after completing the engineering analysis/development phase of the project. Unfortunately, after all the work so far, *the most important thing has yet to be done for the clients.* They must have answers to two imperative questions:

- Will it do what I need done?
- Can I afford to do it?

The clients can live with all sorts of exotic engineering alternatives if they get appropriate answers. The bottom line for them is *being able to justify the money spent.*

Therefore, it is imperative that the design engineer develop a comparison of the estimated costs of doing each of the several feasible alternatives to the problem situation. The clients should be presented with a summary comparing the costs and capabilities/attributes of each of the possible "ways to go." The design engineer is *not* obligated to present *only* the alternative with the lowest cost. It's the clients' responsibility to decide whether cost is more important than function or function more important than cost.

Reporting to the clients. The Engineering Design/Problem Solving Process is an open-ended activity. Since every situation involves an infinite number of possibilities, one could work on the problem forever. Eventually, either because of sheer exhaustion and frustration, or because the clients can't wait any longer, the design engineer has to "wrap it up" and "hand it over."

Everyone has their own style and personal techniques for communicating. Books, courses, and entire degree programs are available to learn how to do it. Over and above all the ways one can do it, the reporting should include these three:

- A brief summary of the findings and/or recommendations, in the form of a one-page letter of transmittal or memo to the clients.
- A detailed written report of all that was done with documentation of the recommended alternatives and rationale.
- A verbal presentation to tell the clients what their options are, the consequences of their choices, and to persuade them that they should either go with it or buy an additional study of other alternatives.

There are some imperative objectives that must be met in reporting any design solution to the clients:

- **Credibility.** The presentations (both written and oral) must look and sound like they were prepared by competent engineers who knew what they were doing, who did a thorough job of researching and analyzing, and who have covered all the bases. The credibility of the study and those conducting it must be established and verified or the clients will be scared out of making any decision except to run to someone else for help.
- **Explanation.** The presentation must explain what is really needed and why the following solutions and/or alternatives are being presented.
- **Thoroughness.** The clients must know how vigorous and extensive the study was and that one or more of the alternatives being recommended are survivors of a lot of rejected

alternatives. The clients need to know that what they're getting is about all they can get under the given circumstances. The project has failed if the clients develop the slightest suspicion that the study missed some rather obvious or intuitively possible alternatives. The clients should be convinced that such a thorough study was made that there is little or no chance of finding a better way.

- **A clear answer.** The report should give the clients a clear path to their decision. It either should strongly recommend a preferred solution compared to lesser alternatives, or it should present more than one option with clearly defined cost/benefit summaries. A good design report should always compare more than one alternative.

WELL, THEN . . .

All this is to say that engineering design is nearly always the business of solving a client's problem. Thus, a design engineer is a staff-level consultant to management. The outcome of a design engineer's work is sound technical advice and clearly defined alternatives upon which an administrative judgment and decision can be made. The design engineer's problem solving procedure must, therefore, emphasize both the creative process of developing feasible alternatives and the necessity of advising the client of the cost/benefit of the options.

SELECTED REFERENCES

1. Leech, D. J. *Management of Engineering Design.* Wiley, 1972.
2. Love, S. F. *Mastery and Management of Time.* Prentice-Hall, 1978.
3. Martin, C. C. *Project Management: How to Make It Work.* American Management Association, 1976.

4

On Being Creative . . . Even Innovative and Clever

The creative ability of human beings could easily be regarded as our greatest natural resource and, no doubt, our greatest *undeveloped* natural resource. It is this human quality that makes the impossible real, anticipates the future, and extrapolates reality. It is the principal source of our progress, our most distinguishing ability—a noble attribute of the human mind.

The challenges brought on by increasing awareness of the potential of nature have created healthy respect and a pressing demand for creative efforts. Within the last quarter of a century, the engineering profession has been forced to rely heavily on its more creative members. Now we find that our dependency on the creative design engineer is placing a premium on creative ability.

The age of miracles has created an insatiable demand for the creative and unusual solution—the innovative approach—fresh ideas—unique concepts. Product innovation has increasingly become the difference between success and failure. Many companies owe their entire production to products that did not exist 10 years ago.

The competition created by product innovation has forced the establishment of industrial research and development centers ("idea factories") to stay abreast of technological progress. To staff these centers, recruiters looked for creative people with ideas—inventors who could create new designs and develop new products that would create totally new markets. Then, of course, there were the inevitable questions: Who are these people? How would I recognize them if I met them? How do they get that way? Can we take any engineering types and train them to be that way? How does a creative person create? What do you mean when you say you want a creative engineer? What in the world is creativity in the first place?

THE LEAST UNDERSTOOD PHENOMENON

While we don't understand creativity at all, we have observed it long enough at least to generalize about it. As one writer put it, "It's the manifestation of a fundamental ability to relate pre-

It's a spontaneous, irrational, intuitive process of imagination—a mysterious mixing of previous experiences and combinations that are totally new.

viously unrelated things." It is the ability to conjure up images and ideas that are beyond sensory experience—to extrapolate conventionality into an imaginary world. It is a spontaneous, irrational, intuitive process of imagaination—a mysterious mixing of previous experiences into patterns and combinations that are totally new and unique.

But how does it happen? Is it just accidental hallucination—random fallout from memory? Is it a genetic accident—a rare talent—that you either have or you don't, like artistic or musical ability? Or is it something everyone has, but only some have learned to use? Is creativity a learned ability like being able to read, memorize, etc.?

WELL, WHAT ARE WE DOING ABOUT IT?

According to Taylor (13), creativity is the one area of behavioral science that has been given the highest research priority. And indeed a great deal has already been done and much has been learned. As with most of the other characteristics of the mind, we have yet to understand the "why" and the "how" of creativity. But a wealth of information (8, 11), with considerable practical significance, is being published on the subject of creative ability. Notably:

- We have developed techniques for measuring relative creative ability.
- We have discovered that creativity is not an inherited uniqueness. We all have the ability in varying degrees—and it can be cultivated.
- We have examined the backgrounds of creative people to determine the influence of their environment and found it to be significant.
- We have studied naturally creative people to learn how they act and react and have identified their personality characteristics.
- We have developed some useful techniques for teaching people how to become more creative.
- We are studying children and have learned some ways in which to capitalize on their native creative ability and to prevent its decay.

- We are investigating the relationship of intelligence to creativity and have discovered a disturbing paradox—they don't correlate.
- We are developing insights into the utilization of creative talent and the prerequisites of a fertile working environment to enhance and optimize the production of creative people.

MISAPPREHENSIONS

"Aw, come on now! You mean to tell me you can teach anyone to be creative?" One of the most common misconceptions about creativity is that it is a talent you are born with, and that you either have it or you don't. This misconception is not entirely unfounded. Our most creative people were indeed born that way. They were endowed with a fortuitous and rare combination of attributes that produced a highly developed natural creative aptitude. They are the people who become our most eminent musicians, artists, writers, poets, and scholars. These one-in-a-million genetic rarities are shining examples of the natural potential of people. Although it cannot be denied that those who have been blessed with great natural creative ability are rare and unique, it has been firmly established that all intelligent people (meaning IQ $>$ 100) possess considerable creative potential.

We know that most mental attributes can be learned. For example, some people have amazing *natural* photographic memories. Although we all have the ability to memorize, most of us cannot match the ability of these people. But we have learned how to teach people to improve their ability to memorize. It is now quite possible to train a person to compete with the feats of people with naturally photographic minds.

Creativity is now regarded as a mental ability that may be improved by training. It is also recognized that people have varying capacities to learn. It would be naive to assume that everyone, with proper training, could become an Einstein, Beethoven, Rembrandt, or Shakespeare. Similarly, few of us could be trained to be great operatic singers although we all can sing(?). But, even the worst singer could be improved considerably with training. So it is with creativity—even the least creative can improve his or her ability by learning to override the psychological factors that inhibit it.

Another common fallacy is that only the most intelligent are creative—that if you make only ordinary grades you could never

THE HIGHLY CREATIVE PERSON

Childhood Characteristics (5):

- Persistent—purposeful
- Quickly thinks of alternatives
- Sees gaps—finds hidden meanings
- Self-winding—self-feeling
- Toys with ideas
- Accepts disorder
- Tremendous energy
- Lags in verbal ability
- Attracted to the mysterious
- Playful—spirited in disagreement
- Emotionally sensitive
- Finds fault
- Courageous—adventurous
- Takes risks

Adult Characteristics (8):

- Seeks autonomy and privacy
- Dedicated to problem-solving tasks
- Aggressive in goals sought
- Relentless worker—great zeal
- Does not value job security
- Likes to clown around—childish play
- Unimpressed by status symbols
- Rejects theological arguments
- Good sense of humor
- Intelligent (IQ: 100–140)
- Likes supervision—orderly organization
- Accepts chaos and change, anti-symmetry
- Insensitive to others' feelings
- Likes to explore ideas
- Nonconformist—enjoys non-conformity
- Independent—observant—says what she or he thinks
- Gullible—open to experience
- Can easily accept failure
- Needs continual reinforcement—an understanding listener

excel creatively. This has not been found to be the case. There is no direct correlation between our measures of intelligence and creativity. Many people who demonstrate a high level of creative output have poor scholastic records. Although IQ among creative people varies considerably, it is above average (ranging from 100–140). Creativity is rare among low-IQ or very high-IQ people. Further, there is considerable evidence that creative ability is strongly influenced by environmental stimulation, i.e., what happens to you while you are growing up.

> As long as you believe you can't—you won't!

Finally, another popular notion is the common stereotype of creative people that casts them as sloppy undisciplined Bohemians or as emotionally unstable and effeminate. When people matching this stereotype were measured for creative potential, it was found that most had very little creative ability. MacKennon (6) reports that the truly original and creative engineers are deliberate, reserved, industrious, and thorough. They include introverted and extroverted thinkers, are self-assertive and dominant and have a high level of energy.

HOW CREATIVE CAN YOU GET?

Creativity expresses itself at several levels, according to Taylor (13). From the lowest to the highest, these levels are defined as:

- **Expressive**–spontaneous
- **Productive**–heightened realism
- **Inventive**–ingenuity
- **Innovative**–development by modification
- **Emergentive**–formulation of new principles.

Children function at the *expressive* level of creativity. Their creative acts are impulsive–off the top of the head, with no care for skill, function, criticism, or any sort of restraint. You ask children to draw a picture–and they draw one. They don't ask what kind you want or concern themselves with whether they can draw a picture or not–they just do it–good, bad, or otherwise. It is this freewheeling level of creativity that is the fountainhead of every person's creative potential. And, it is an attribute that is generally destroyed during the maturing process.

Most engineers are called upon to express their creative abilities at the **productive** and **inventive** levels. Productive creativity involves the creation of practical combinations to fulfill a preconceived plan–achieving realism. This area of engineering is design detailing. **Inventive creativity** is a higher level of ingenuity in the development of new ways to do old things–inventing an automatic machine to replace a hand operation, for example.

Creative design engineers involved in conceptual engineering activity are expressing a level of **innovative creativity**. This is a highly sophisticated creative aptitude. It requires creative insight to devise new meanings and expressions from existing principles and theories.

This is the research and development area of engineering that creates products that have no precedent from new scientific discoveries. Engineers who can function at the innovative and inventive levels are the most sought after creative talents in engineering.

The highest—**emergentive**—level of creativity transcends all that's known. It evolves entirely new concepts and principles to extend our awareness. This is the realm of the rare and gifted creative genius—our most outstanding scholars, artists, and musicians.

It is this very small number of people who have made the most notable contributions and who have become the stereotype of creative aptitude. Because of the monumental influence of these few, people in general tend to discount and discredit the existence of the extremely necessary and valuable lower levels of creative aptitude.

Highly creative people are at work in all these levels. The value of creative activity is not identified by how abstract or obtuse (odd ball) it is, or by how much it affects the total scheme of things. All levels of creativity are valued in any area of application. It is the most important parameter in all human activity. Without creative effort in every human pursuit, we would still be animals living in caves.

HOW DOES A CREATIVE PERSON WORK?

When a person responds to a situation requiring creative effort, there is a general pattern to the creative process. We have no idea why it is this way, but the process is universally experienced. The process evolves with varying degrees of emphasis in the following sequence:

- **Preparation**—exploring the situation, seeking implications, defining the underlying concepts.
- **Search**—collecting facts, defining constraints and possible outcomes.
- **Ideation**—searching for alternatives.
- **Frustration and incubation**—mental struggle and processing.
- **Illumination**—perceiving the most promising way.

- **Evaluation**—testing for goodness of fit.
- **Execution**—developing details and potential.
- **Communication**—defining, describing, and seeking acceptance.

As this sequence occurs, creative people perceive a great number of related facts, incidents, similarities, and analogies. They are highly associative and nonconventional. They log a tremendous variety of relationships associated with the problem. Their attitude allows them to mix their storage of knowledge freely, without the constraints of conventional prejudices. This ability is sometimes referred to as "plastic perception."

It is a lonely and individual process. There is a period of struggle and frustration leading eventually to what Gordon (3) refers to as the Hedonic Response. This is a feeling of elation—an intuitive perception that an idea is "right"—an intense feeling of satisfaction and achievement—a flush of inspiration. These moments of revelation, when the "light flashes," usually occur in periods of relaxation after a long struggle with the concepts of the problem—after the subconscious mind has had time to react to the pressures of the struggle.

This subconscious "computing" phase of "inspiration getting" is least understood. However, the "programming" or "stage-setting" phase of the creative process is becoming a well organized art. Osborn (9) was one of the first to develop a systematic procedure for the creative problem-solving process. Gordon (3), Allen (1), and others have enriched the techniques. In the next chapter, we will discuss the practical approaches to getting ideas.

Creative people are self-starting—self-motivated to respond creatively. They have a compelling urge to seek new insights. They are driven by a constructive dissatisfaction for what exists, and strive to achieve the "elation of revelation." They are "idea-holics" on a continuous inspirational jag. The rest of us, who aren't hooked, may occasionally become involved in a creative orgy, but because of our inexperience, suffer from hangovers of frustration and depression. However, with determined and steadfast imbibing of the creative spirit, we too can become idea-holics who can think any veteran under the table.

IDEA-HOLICS ANONYMOUS

Believe it or not, there is a widespread temperance movement dedicated to stamping out creativity. The independent, nonconforming, unorthodox thinking of highly creative people inherently generates conflict with those people they deal with. This conflict, and the resulting prejudice of teachers, classmates, employers, and even parents, begins early in life and continues throughout one's career.

Highly creative children are a bundle of energy; they are extremely perceptive, curious, aggressive, full of devilment and play, and get into everything. They have a personality that is a constant source of irritation and embarrassment to some parents. Their boisterous, playful, human and perceptive inquisition is regarded as mockery, rudeness, and disrespect by adults. Their energetic spirit and adventuresome curiosity are too often interpreted as disobedience and disregard for discipline and authority.

A group of secondary and primary teachers were asked to list the traits they most desired and least desired in their students (4). All of the "desirable" traits matched the personality of people with high IQ's—promptness, courteousness, popularity, receptivity, good memory, etc. The "least desirable" traits (unwillingness to accept say-so, courageousness, adventurousness, inquisitiveness, willingness to take risks, visionary thinking, etc.) are the characteristics of markedly creative people.

Employers find that the personality of highly creative individuals is a disruptive influence among employees. Their aggressiveness, disdain for administrative organization, and continuous crusading for unconventional ideas all create conflict. The conventional people in our society are basically prejudiced against a nonconformist, jealous of someone with bright ideas, and dislike unorthodox behavior (such as aggressiveness or a domineering attitude). People react against a person who takes independent action and won't go along with the crowd. They tend to distrust someone who has wild and impractical ideas. It is fortunate indeed that the creative personalities are protected by a high ego strength and are not socially oriented, or the knot-headed prejudices of society would surely have destroyed them. Unfortunately, however, for every highly creative person, there are thousands whose creative potential has been virtually wiped out by the degrading influences of the social environment.

CREATIVITY WATCHING

When you watch creative people doing their thing, they:

- Will always be busy at their own business.
- Prefer to dress differently.
- Do more than was required, assigned, or expected.
- Are often preoccupied with thinking (really).
- Get intrigued and involved using simple things in clever ways.
- Make up games in group situations.
- Come up with ingenious methods of doing it differently than instructed.
- Draw elaborate pictures, intricate designs, and dramatic doodles while listening.
- Talk enthusiastically about their discoveries, ideas, and inventions.
- Experiment with and explore familiar objects.
- Question beyond merely "Why?" and "How?"
- Delight in trying something new.
- Watch it, listen to it, smell it, taste it, and feel it.
- Don't mind (or even realize) that they are "different."

Paraphrased from "Non-Test Ways of Identifying the Creatively Gifted," by E. Paul Torrance (4).

WANNA KNOW WHY YOU AREN'T CREATIVE?

Because you're scared! From the time you first started to use your imagination and explore what you could do with it, you have been ridiculed, punished, teased, embarrassed, made fun of, laughed at, bawled out, intimidated, threatened, derided, and ignored. Since your hide isn't a foot thick, and you like to get along with people, over the years you have learned to be very careful about following a creative impulse that could cause a lot of grief for yourself. You are real gun-shy in creative situations.

What are you scared of? According to McPherson (7) you're scared of:

- making a mistake
- making a fool of yourself
- being criticized—especially by superiors

- being too pushy—of crusading
- having your ideas stolen
- saying no to everyone who wants help
- being in the minority
- being different—not conforming
- taking time to engage in fantasy
- not knowing enough about the situation.

Wanna know why you aren't creative?

All of these fears are the result of conditioning—of being disciplined to conform to social patterns—of learning to avoid conflict.

In addition to all these emotional blocks to your creative ability, you have managed to pick up some perceptual blocks. For example:

- You tend to chicken out of an idea situation because you feel the technical area is over your head.
- You tend to narrow the problem too much—instead of thinking "big" or thinking "way out."

- You can't help wanting to be practical.
- You have difficulty in investigating the "obvious"—you are always looking for a gimmick.
- You invariably will be drawn to dealing with the effect rather than the cause.
- You have difficulty focusing on what needs to be done—you tend to want to look at what someone else (the boss) wants.

Well, what does it all boil down to? When you think about it, you can't be creative and be a "good ol' boy." The more you try to be accepted, the more dependent you become—and the less inclined you are to think independently.

NEGATIVITY
Attributes of the Uncreative
- Resistance to change
- Desire for conformity
- Competitive jealousy
- Desire for security
- Fear of ridicule
- Cynicism
- Concern for effect rather than cause
- Distrust of wild ideas
- Fear of failure
- Desire for organized routine and order
- No desire to experiment

You cannot be creative and conform to social pressures. You have to be independent and objective. You have to say "no" often enough so that you will have time to devote to creative effort. You have to develop enough courage to "push" an idea—in the face of all kinds of negative reaction. Which means you have to learn to ignore what "people" will think. Obviously, it's a lot harder to be creative than to go along with the crowd and be liked.

THE LEAST YOU CAN DO . . .

We're learning how to bring out a person's creative potential. There are three very successful techniques: Osborn's *Applied Imagination* (9), Allen's *Morphological Creativity* (1), and Gordon's

Synectics (3) are three desirable methods for enhancing one's ability to think creatively. For design engineers this is a must. But regardless of whether you are personally involved in the creative problem-solving activity, you will invariably be in a position to have a direct influence on the creativity of others—most probably as a parent, and quite likely as a supervisor of design engineers.

Praise the ideators! (It's the best thing you can do for them.)

The least you can do is try to help creative people develop and use their potential. Lagemann (5) suggests several do's and don'ts that will go a long way to encourage and develop the creative potential of children. For example:

- **Don't** discourage fantasy.
- **Don't** hold them back.
- **Don't** shield them from mistakes and failures.
- **Don't** impose sexual stereotypes ("sissy," "girls don't do that").
- **Don't** judge them by their reading or writing (creative children are notoriously slow here).
- **Don't** pin them down and make them have a logical reason for everything.
- **Don't** subject them to ridicule.
- **Do** reward them for creative acts—hand out encouragement when they do something creative.
- **Do** help them get along with people—show them how to be tolerant of other's ideas, assertive but not domineering, honest but not critical.
- **Do** help them accept their minority—unpopular—status among other children.
- **Do** encourage them to explore and experiment.*

There are going to be more than enough thoughtless, stupid people who will discourage your children—they ought to have at least two staunch allies in their own parents!

*J. K. Lagemann, "How We Discourage Creative Children," quoted from the March 1963 issue of *Redbook Magazine,* Copyright © 1963 by McCall Corporation.

You can also be a great deal of help to your engineering colleagues. Anyone who is trying to create needs a "friend"—especially

THE THREE P'S OF CREATIVE PROBLEM SOLVING

- **Pursuit**—making it more important to do than a lot of other things you could be doing instead.
- **Perseverance**—being willing to put up with the unsolved situation for a long time.
- **Persnicketiness**—being dissatisfied with what you know about it and what you could do about it—being compelled to find a solution that satisfies you.

a "friendly" supervisor. McPherson (7) points out that a healthy working environment for creative engineers depends on a supervisor who will:

- encourage free and informal communication.
- discourage crash deadlines.
- make allowances for failures.
- listen without prejudice.
- reward unusual thinking and ideas.
- give recognition that ideas are being understood and used.
- pair the creative people with sympathetic co-workers who can stimulate them, can interact with enthusiasm, can pick up the ideas and put them into use.

Creative people have to be listened to. They need a sounding board—someone who can react with enthusiasm and who can work to see that the ideas develop into useful products. Creative people in general lose interest in the details required to develop a concept. They prefer, instead, to tackle the next problem. But they need to have the satisfaction that their ideas are being used and are contributing. It is the payoff that encourages them to forge ahead.

Creative people invest a great deal of themselves in the development of a concept. It is a long, difficult, agonizing mental struggle and understandably develops a great amount of pride and enthusiasm for the idea. They are bursting to get it going and can't wait to get everyone jumping up and down about it. They are virtually wiped out if there is no one in a decision-making or expediting position who is sympathetic to the idea.

SELECTED REFERENCES

1. Allen, M. S. *Morphological Creativity.* Prentice-Hall, 1962.
2. Buhl, H. R. *Creative Engineering Design.* Iowa State University Press, 1962.
3. Gordon, W. J. J. *Synectics.* Harper & Row, 1961.
4. Gowan, J. C., Demos, G. D., & Torrance, E. P. *Creativity: Its Educational Implications.* Wiley, 1967.
5. Lagemann, J. K. "How We Discourage Creative Children." *Redbook Magazine,* Mar. 1963, p. 44.
6. MacKennon, D. W. "The Creative Worker in Engineering." Paper presented at the 11th Annual Industrial Engineering Institute, University of California, Berkeley, 1959.
7. McPherson, J.. H. "Are You Creative?" *Product Engineering,* Nov. 17, 1958, p. 28.
8. McPherson, J. H. "The Relationship of the Individual to the Creative Process in the Management Environment." *ASME Paper* 64MD12, May 1964.
9. Osborn, A. F. *Applied Imagination.* Scribner's, 1963 (contains 409 references).
10. Raudsepp, E. "Removing Barriers to Creativity." *Machine Design,* May 24, 1962, pp. 138–143.
11. Stein, M. I. "Survey of the Psychological Literature in the Area of Creativity." *Research Center for Human Relations Report,* New York University, 1962.
12. Taylor, C. W., & Ellison, R. L. "Biographical Information and the Prediction of Multiple Criteria of Success in Science." *NASA Report,* Project NASA-105, Aug. 1963.
13. Taylor, I. A. "The Nature of the Creative Process." *Creativity,* Hastings House, 1959, pp. 51–82.

5

Ideation . . . Seeking Alternatives

One of the most valued attributes of the engineer is the ability to create ideas. Ideas are all that keep an engineer from being replaced by a computer or being reduced to doing routine tasks on command. Since routine tasks by command are precisely what a computer is made for, it can beat out a human hands-down in speed, accuracy, and cost. In fact, the business of getting ideas (being creative) is taking on survival proportions in engineering. It's a last frontier—the one thing engineers can do that hasn't been mechanized.

Ideation is the mental process of stimulating the imagination to produce concepts and ideas in order to solve a problem at hand. We've gained some insight into how creative people stimulate their imagination, and we have developed an "art of idea-getting." It's analogous to brain programming, and it works. Osborn calls it "applied imagination," Allen calls it "morphological creativity," Gordon calls it "synectics." Each approaches the matter in a different way. To get in on this survival training, you need to gain some insight on what ideas are, how they come about, and what you have to do to stimulate good ones.

THAT'S WHAT IDEAS ARE MADE OF . . .

What a fantastic proposition it would be if we could take a pill to stimulate our imaginations to the peak of our creative potential. If we had such pills, we would undoubtedly guard them with the fear we attach to the atomic bomb. It is doubtful that the world would accommodate such an avalanche of revelation. It could be too much of a good thing—like being smothered in whipped cream.

We may not be too far from discovering such a stimulus. There are Indians who live in the jungles of Mexico who have for generations used a certain mushroom that creates hallucinations. They have (wisely) evolved a rigidly controlled religious ritual around these hallucination orgies. Scientists who have experimented with this stimulus have reported being taken for a horrifying night ride on the wildest mare they have ever seen. Overdoses of "pep" pills are also reported to create hallucinations of frightful proportions.

These chemically induced hallucinations seem to be the manifestation of overstimulation of the idea-mechanism or the imagination process. A totally uncontrolled scrambling of all the memory storage results, which produces a sequence of images that almost overwhelms the rational mind (at least gives it a dreadful scare).

Ideas are images produced by the stimulus of the moment. They are combinations of one or more previous sensory experiences which have been indexed in the memory by some sort of reference key. Every sensory input that has been received since you were born has been logged in your memory and is theoretically available for recall. (Can you imagine how many miles of computer-tape storage it would take to record the sensory inputs of the conscious lifetime of a single human being? One of the greatest mysteries of our universe is how our brain stores all this information in the space it has.)

When you seek an idea, you program your brain to seek out all past recordings of experiences related to this situation. Since there are many, each with varying levels of pertinence, images are formed from random mixes of these stored memories. Since no two memory storages are identical, no two people will respond to the same stimulus in the same way. There is seemingly no limit to the variety of idea responses that can be achieved for a given demand situation.

Obviously, to have ideas you have to have memorized or recorded experiences to draw upon. The more creative people are,

It's our last frontier—the one thing we can do that we haven't been able to mechanize.

the more observant they are and the more they relate the present to previously observed experiences. When highly creative people observe something, they are quick to spot similarities to other things, note analogies, associate other events to it, and note the relationship of the environment to it. When the experience is logged in the memory it is tagged and cross referenced under a variety of "call outs"—reference keys. Later, when another situation comes up that has any one of the reference keys that were associated with the remembered experience, it may be recalled and become an ingredient for an idea.

There are many sensory experiences that a person has had that are never consciously identified by some sort of a reference key. These events are rarely recalled by conscious effort. There are occasions though when such an event will be recalled for no apparent reason. We all have had occasions when we suddenly remember a remote incident in the past that we hadn't thought of in years. These reminders seem to occur when they have no apparent connection to the current sensory event. The reference key that triggered the recall may have involved a combination of sights, sounds, and smells that we associated with the recalled event long ago.

IMPULSE IDEATION

There is a vast difference between spontaneous recall and getting ideas. One is nostalgic; the other is a source of action—an answer to a question. Most of us as we go about our daily routines occasionally get an impulsive idea—a little brainstorm. An exciting little bit of inspiration will come to light to solve some problem that we have noted at one time or another but for which we have no particular concern. We also use our creative ability impulsively in frivolous, party situations—when playing pranks or teasing someone—when there is no restriction on how ridiculous we can let our imaginations get.

Impulsive ideation is generally all in fun and strictly reactive to the spirit of the moment. Everyone is pretty adept at this type of mental play. It's exciting and stimulating; there is no pressure to produce an idea and no particular demand for any specific kind of idea, either.

DEMAND IDEATION

The most difficult and useful creative activity is demand ideation—generating ideas to solve a particular situation in which you have had no prior personal interest. You are asked to find a way to solve a problem. Generally, you have no immediate notion of what to do. You most probably will not have had any previous experience in the situation; for all practical purposes, you are caught stone cold.

This is the area of creative problem-solving and it is the crux of engineering design. Few of us are subjected to demand-ideation situations during our childhood. Consequently, most of us (with the exception of the naturally creative) cannot respond to the demand-ideation situation because we have never had to, and don't know how. When such a problem is posed, the best we can do is to respond impulsively (relying on the bolt of lightning) if at all—then we are faked-out in futile frustration ("Wow, that's a tough problem"). Creative people, on the other hand, have learned how to approach the problem so that they can cut it up into bits of information that will stimulate their imaginations to respond in a great number of ways. It is this phase of the creative process that is teachable. It is an approach people can learn to use which will greatly improve their creative ability.

There are several different demand-ideation techniques that are very effective idea stimulators. Each technique is a supplement to the other. For fruitful ideation, you should employ them all. Each stimulates the imagination in a different way and opens up new avenues to unique and unusual ideas. These ideation techniques are:

- The trigger-word technique
- The checklist technique
- The morphological chart technique
- The attribute-seeking technique
- The Synectics technique
- The brainstorming technique

No single technique is superior to another, nor is any one technique applicable only to specific situations. To capitalize effectively on your creative potential, you need to learn to use all of them.

THE TRIGGER-WORD TECHNIQUE

To stimulate ideas, you have to ask yourself pertinent questions to define the problem. Each question you can ask represents another way of looking at the problem. Each different way you can phrase the problem stimulates the imagination to associate related ideas to the situation.

The trigger word is nothing other than the verb in the problem definition that you are using at the moment. Consider the following simple situation: We are given an object at position A. We want to MOVE this object and locate it at a position B after some elapsed time interval. The trigger word here is MOVE.

We can define this problem in a variety of ways. Each new definition, however, is obtained by substituting another verb in the problem statement. For example: How can we PUSH the object from position A to position B? Or, using other trigger

"TRIGGER-WORD" SAMPLES

Move		
cartwheel	push	glide
float	pull	slide
sling	shove	tumble
throw	roll	lift
swing	fly	drop
whirl	leap	drag
kick	run	slip
Get Open		
explode	melt	swing
tear	bend	roll
cut	peel	cleave
slice	unwind	whirl
rip	unravel	twist
screw	fling	
Put In		
squirt	squeeze	fall
drop	swing	knock
pour	construct	tilt
shake	reconstitute	turn
slide	dump	fold
funnel	trickle	ram
Get Shut		
bend	roll	twist
fold	solder	screw
wrap	lock	rivet
weld	clamp	glue
bolt	enclose	mold
close	seal	pin
clip	nail	press

verbs: How can we FLOAT it? FLY it? PULL it? HAVE it at B? ROLL it? Every different trigger verb has a different set of connotations and creates a new batch of ideas for ways to solve the problem. The word PUSH will recall some previous experience associated with pushing. You may see a bulldozer, a hydraulic ram, or a group of people doing the pushing.

When you have exhausted your imagination for pushing situations, try the trigger word PULL. This word recalls a team of horses, a cable and winch, etc. And so it goes, using alternatives for the action verb in the problem statement. It is a methodical method for increasing the scope of idea stimulation. It is a simple method for forcing a variety of ways to conceive the nature of the problem. The greater the variety, the more ideas you will stimulate.

THE CHECKLIST TECHNIQUE

The checklist serves to remind you of other ways of looking at the problem so that new imagination stimuli can be introduced into the ideation process. There is no standard list needed. Any list of alternative questions you can ask yourself about the problem will serve the purpose. Osborn's checklist (6) is widely used as a model since it includes a broad grasp of the fundamental variations in any problem. He lists the following:

> Mini-, Magni-, Modify,
> Simpli-, Satis-, Stupify,
> Associ-, Negoti-, Eliminate.
> That's the way to ideate!

- Put to other uses?
- Adapt? (like, suggest, copy, parallel, etc.)
- Modify? (change, color, motion, order, taste, shape)
- Magnify? (stronger, higher, longer, add, extra, exaggerate, etc.)
- Minify? (subtract, condense, lower, shrink, cut off, etc.)
- Substitute? (what else, other, replace, etc.)
- Rearrange? (interchange, alter sequence, scramble, etc.)
- Reverse? (opposite, backward, upside down, inside out)
- Combine? (blend, assortment, hookup with, etc.)*

*Reprinted by permission from A. F. Osborn, *Applied Imagination,* Charles Scribner's Sons, 1963.

Each of these questions suggests a series of cognate questions (in parentheses) that should stimulate your imagination to explore the less obvious concepts surrounding your problem. They serve to force you to explore notions that are unusual and quite impractical on the surface. But each avenue of attack provides a fertile new ground for your imagination to work in. For instance, the question "Reverse?" is a powerful stimulus to abandon conventional paths of thinking and consider more improbable concepts.

IDEA STIMULATORS

- Look for similarities.
- Look for associated equipment and conditions in the vicinity.
- Look for contrast:

Exaggerate	Reverse
Magnify	Switch
Minify	Rearrange
Modify	Opposites

- Look for inversions:

Backwards	Inside out
Upside down	Mirror image
Change places	Left for right
Stop the moving–move the stationary	

Reverse implies inversion. To invert can connote letting the moving element be fixed and allowing the fixed parts to move. The Germans in World War I built a radial aircraft engine by the inversion technique–the propellor was fastened to the cylinder block that rotated about the crankshaft, which was rigidly fastened to the frame of the plane. The engine has some inherent undesirable features, but it nevertheless was a unique approach.

THE MORPHOLOGICAL CHART TECHNIQUE

This technique has been developed in considerable detail by M. S. Allen (2). The method involves analyzing the problem to determine the several independent parameters that are involved in the problem. Each of these parameters is then considered independently for possible alternative methods. All are tabulated in a matrix which can be cross correlated to produce hundreds or thousands of idea solutions to the problem.

PARAMETERS Design Characteristics and Requirements	IDEAS Design Alternatives						
Input Motion	*Rotating	Oscillating	Linear	Reciprocating	Etc.		
Input Source	One Hand	*Two Hands	One Foot	Two Feet	Hand & Foot	Etc.	
Input Device	*Crank	Turnstile	Pedals	Lever	Treadmill	Etc.	
Output	Propeller	*Paddle	Paddle Wheel	Fin	Screw	Jet	Etc.
Mechanism	Gears	Chains	*Belts	Pump	Linkeage	Piston	Etc.
Operator Position	Sitting	*Standing	Kneeling	Straddling	Etc.		

SAMPLE MORPHOLOGICAL IDEATION CHART

Problem: Design a Manual Propulsion System for a Small Boat

The independent parameters are listed in a column and each idea for achieving the parameter is listed in rows. This forms a grid. Every combination of parameters and ideas that you can get across this grid represents a possible solution.

The morphological procedure illustrated here is a sample chart of possible solutions for a manually propelled propulsion system for a small boat. In the left-hand column are listed some of the requirements and characteristics of the design, such as input motion, the various input sources for this motion, input devices, output mechanisms, operator positions, etc. The list shown in the left-hand column can be varied in number and type of parameters, depending on the characteristics and requirements you attach to the design.

Each row lists a series of alternative ideas. For example, what are the kinds of input motions that could be used to propel a boat? Well, we could have rotating motion, oscillating motion, linear motion, reciprocating motion, and others. The same procedure is used for each of the other parameter requirements listed in the left-hand column. If we should combine one idea from each row with one idea from each of the other rows, we would have a definition of a manual propulsion system for the boat. For instance, if we combine the ideas marked by asterisks, we obtain a system that is operated by a standing operator who uses both hands to turn a crank mechanism connected by a belt drive to a linkage mechanism that operates a paddle.

The number of combinations of idea concepts can be very large. For example, four rows, of ten ideas each, produce 10^4 (or 10,000) possible solution concepts. This chart is an extremely powerful device because it methodically and categorically forces you to ideate various aspects of the problem in considerable detail.

Any of these idea-forcing techniques are obviously going to bring forth many ridiculous and impractical ideas. However, every idea is valuable if for no other purpose than to stimulate the imagination even further. A totally ridiculous idea can ultimately trigger an extremely original, unique, and valuable new concept.

THE ATTRIBUTE-SEEKING TECHNIQUE

The attribute-seeking technique is in some respects similar to the morphological approach. In this case, however, instead of listing the independent parameters, you single out the basic attributes of the problem situation for individual ideation. That is, the attributes are the essential requirements of the problem where a decision between alternatives must be made. In the boat-propulsion problem the required drive has to have (1) manual power, (2) a means for causing the hull to move, and (3) a power-conversion device. For each, you have to decide on one from all the possible alternatives for achieving that particular essential requirement.

ATTRIBUTE LISTING

Devise a potato-chip bagging machine . . .

Attributes:

- Supply of chips
- Measure of chips
- Bagging of chips
- Removal of bags
- Production rates
- Cleanliness—sanitation—accuracy

Devise a new study lamp . . .

Attributes:

- Light distribution—work area
- Light distribution—nonwork area
- Light-producing device
- Light-source support structure
- Start-stop device
- Power supply

Devise a new lawn mower . . .

Attributes:

- Grass-blade cutting
- Propulsion over lawn surface
- Operator control
- Power supply
- Trimmings disposal

Obviously, attributes are subject to a wide range of interpretation. Each of us would probably devise a different list. It makes no difference what the list is—the important thing is that a variety of approaches to the characteristics of the problem is made available for consideration. The search for the alternatives to each independent parameter in the morphological chart, or for each attribute, is an idea-seeking problem in itself. The trigger-word approach and the checklist could and should be employed to expand the list as far as possible.

THE SYNECTICS TECHNIQUE

W. J. J. Gordon (4) has devised a group-ideation technique that basically forces unconventional thinking by "making the familiar strange" through the use of analogies (metaphors). It's essentially a two-stage process:

Stage 1—"Making the strange familiar." A concentrated effort to search out *all* aspects of the stated problem, especially the obvious elements and underlying concepts. The defining process is pursued until all aspects of the situation become so familiar it begins to stimulate comparisons and analogies to seemingly totally unrelated situations, concepts, and events. The key here is to force unconventional understanding by searching for what is *really* going on here in a truly basic sense.

STAGE 1: UNDERLYING CONCEPTS (THE SYNECTICS TECHNIQUE)

Devise a new lawn mower . . .

Underlying concepts:
- Grass cutting
- Uniform height of ground cover
- Neat, pleasant yard appearance
- Grass growth

Devise a new trash collector . . .

Underlying concepts:
- Trash storage
- Trash removal
- Trash accumulation
- Waste products
- Disposal of unwanted items
- Separation for recycling

Devise a new mass-transit system . . .

Underlying concepts:
- Relative locations of living quarters and work
- Movement of people
- Independence of individuals

For example, suppose the problem is to devise an automobile accelerometer—an instrument on the panel to show a sport how much jazz the car has. You might explore, for instance, the underlying concept of inertia, since inertia is one of the fundamental consequences of acceleration. Or, if you are to design a home trash-disposal appliance, you should investigate the underlying concept of waste.

Stage 2—"Making the familiar strange." This is the ideation stage. The power of the technique lies in seeking and defining metaphorical analogies to events and concepts in areas totally out of context with the given problem situation. To force diverse analogies, Gordon suggests four metaphorical mechanisms to assure unconventional thinking:

- **Personal analogy.** What if I personally became the device to accomplish the solution? What would I be doing?
- **Direct analogy.** What is going on in nature or anywhere that seems to involve some of the same activities and outcomes needed for this situation? The trick here is to force the opposite. If the situation is a machine process, look for a direct analogy in living nature. If it is a people situation, look for a thing analogy.
- **Symbolic analogy.** Comparing the given situation to an abstraction from our culture. "This whole situation is like an Indian Rope Trick."
- **Fantasy analogy.** "If your wildest wish could come true, what would you wish for?" It involves deferring the reality of the problem by fantasizing a magical wish resolution.

When you approach the creative ideation process through these avenues of thinking, you are compelled to take an unconventional, totally strange, and unfamiliar view.

DEFINING THE PROBLEM

- Study the background.
- Compile known facts.
- Self-questioning:
  Why is it needed? What happened before? After? Is this the only way? What has to be accomplished? What are the constraints? What affects the operation? What is effected? Will it be hot, cold, wet, dry, hard, soft, rough . . . ? How many? How often? Why?
- *Don't* spend a lot of time searching for examples of how someone else solved similar problems.

The Synectics technique is a group-oriented ideation process. It, like brainstorming, succeeds by capitalizing on the combined reactions and stimuli of several brains responding and interacting. It is particularly valuable in those design situations where all the obvious and conventional ways have already been tried.

THE BRAINSTORMING TECHNIQUE

Brainstorming is one of the most useful group-ideation techniques. It was developed by Osborn for use in the advertising business and subsequently has been given a great amount of publicity. It was mistakenly touted as the key to creativity—the cure-all for solving problems. This claim immediately brought forth the critics, who promptly and "scientifically" dispatched it as a fad and a useless game. In their zeal to destroy the myth that it was the answer to creativity, they preferred to try to eliminate it altogether, rather than recognize that it was another helpful technique in the process of getting useful ideas.

Don't stand there and jeer
And throw up your handza;
That stupid idea
Might start a bonanza!

Group brainstorming has two valuable attributes. First, it forces an atmosphere of what Osborn calls "deferred judgment"—that is, a mood of "Who cares how ridiculous we get!" Second, it's analogous to having six different computers working on the same input.

The success of brainstorming depends on the compounding effect of each person in the group responding to the ideas expressed by others. Each person will react differently to any remark, and a great variety of concepts and ideas will result. It becomes idea play and is a very stimulating session provided that no one in the group attempts to evaluate and judge the ideas as they come forth. The mood of the session is totally destroyed when every inspiration is pounced upon and slaughtered on the spot.

The inspirations and ideas that come forth are spontaneous reactions and obviously haven't been weighed and evaluated for their practicality. Any single idea that comes forth in a group session such as this has its faults. The important thing is that it is an idea—good, bad, or ridiculous. Its only value may be that it triggered another possibility that triggered another that eventually led to a totally unique and valuable method.

Brainstorming sessions usually produce an average of an idea a minute—60 ideas an hour. The chances are very good that most of the ideas in the list will be useless—but there will always be several that will be worth looking into further—and that's all you need.

The rules for a useful brainstorming session are simple:

- Use a group of not more than six or eight individuals who are conversant with the field but not necessarily experts or even acquainted with the problem.
- The person with the problem acts as moderator to define the situation and to provide new interpretations of the problem when there is a lag in inspiration.
- Record everything that's said.
- Maintain a light, humorous atmosphere. Encourage members to think freely and bring out wild ideas.
- Caution every member to refrain from judging each idea. This is very important.
- Stop the session when the group begins to tire (usually after 30 to 40 minutes).

LET YOURSELF GO

Whenever you seek ideas, you are urging your imagination to conjure up a wide variety of unique and valuable methods for accomplishing the problem. Each of the techniques for stimulating ideas encourages you to abandon conventionality, forget practicality for the moment, consider what is possible and even what is ridiculous.

Actually, the key to the development of your creative potential lies in being able to overcome the caution, judgment, conservatism, habit, practicality, and fear of ridicule that define the conventional attributes of maturity. Much of the training and conditioning you

do to enhance your ability to develop creative solutions and generate ideas is nothing more than learning to shuck your inhibitions and let your natural imagination have a go at it.

Dinky thinkies never win On any project they begin!

Make it a point to see how ridiculous an idea you can come up with each time you have an idea-demanding problem to deal with.

Think kooky, think crazy, think funny, Start ideas that bring in the money!

It will stretch your imagination and allow you to broaden your attack on the problem. Let yourself go—you won't crack up.

BONE DRY

Any ideation session is a demanding emotional experience. You get into the swing of things and are propelled by a flush of inspiration. But eventually you run out of ideas. You are emotionally wrung out—bone dry—fatigued and frustrated. None of the ideas you've had gives you any satisfaction as having the potential of a good solution—you still feel compelled to seek other solutions —but you're fed up. Then quit!

Getting off the problem after a good hard go at ideation—forgetting it and doing something else—has a great deal of creative potential. It is usually referred to as the incubation period. Somehow—in some way—the subconscious continues to react to the pressure of needing a solution, even though consciously you are fed up with it. Many people have found some extremely useful ideas have come forth "out of the blue" during these rest periods.

WHEN YOU GET STUCK . . .
Try Kidding Yourself

Everyone, at some time, will run out of ideas and get stopped cold. One way to unlock this idea log jam is to paraphrase your problem to friends in the most ridiculous and preposterous way you can imagine:

Invent a tale about the problem and the predicament you are in. Exaggerate the details of the situation, then invent some ridiculous solutions.

Draw them into the situation so they will play with the spoof too. Before you know it, you will have propelled yourself into a prolific brainstorming session that will not only regenerate your own idea strength but could well produce some unique ideas.

Most of these revelations come forth when you are engaged in some routine task like driving, shaving, mowing the lawn.

Incu-, Incu-, Incubation!
Helps idea formulation.

Even though you may not get "the" solution during an incubation period, you will recover your ego strength and, along with it, some new insight into re-examining the problem. This is where having patience with yourself during the creative process pays off.

If you make yourself lay off when you feel fatigued and depressed, and then hit it again when you feel your confidence reviving, you have overcome a formidable obstacle to successful problem-solving.

NOW WHAT?

O.K.! So you tried! You paid attention to all of our good advice and used some of the idea-getting techniques. You should have a fairly respectable list of alternatives. Now comes the important part–searching through the list to find the ideas that have the greatest potential.

There is a natural tendency to go over the list hurriedly, throwing out ideas right and left as stupid, ridiculous, silly, unworkable, unfit, etc. Osborn cautions that the evaluation of the ideas is an important step, not to be taken too hastily or too lightly. He suggests that you evaluate each idea creatively rather than eliminating ideas by using your overdeveloped ability to find fault and criticize.

Those experienced in problem solving techniques suggest using a decision table to evaluate a list of alternatives. It is especially useful for a group evaluation process and a natural follow-on to a group ideation session (as in brainstorming and synectics). Simply, it involves a list of agreed-upon criteria that must be met by the idea. Each member is polled for an intuitive assessment of the idea's strength/potential for meeting each of the criteria needs. Each member is asked to rate the idea using only three levels of vote: + if it favors, 0 if it is doubtful or neutral, and - if it couldn't meet the requirements. It has been shown that pooling the collective judgments of several people on specifically itemized issues produces a highly reliable assessment of which ideas are worth developing (see Sample Idea Decision Table).

In judging each idea, don't dismiss it for what is wrong with it. Look at the idea for what is good about it and for its *possibilities* for refinement. Every idea has its obvious shortcomings. What is really important is the idea's *potential* for meeting each of the criteria. Is it modifiable? Can some changes be made? Do its plusses make working on its negatives worth the effort?

SAMPLE IDEA DECISION TABLE

Need Criteria	Idea # 1	Idea # 2	Idea # 3 . . . Idea # N
Simple	+	–	+
Cost	0	–	+
Lightweight	+	+	0
Safe	+	–	+
Small	–	+	+
High Accuracy	+	+	–
Reliable	0	0	+
Whatever	?	?	?

Summary: It looks like Idea # 3 has a lot of potential but would need some further brainstorming to resolve accuracy and weight concerns.

This is creative evaluation! Here is where you really capitalize on the products of your imagination. You have a long list of possibilities to work with—a whole pile of goodies. Surely with some open-headed, clear-eyed manipulation of all these ingredients, you can evolve some very useful possibilities.

The Creative Process does not begin and end with the listing of ideas. It's just as important that you use your imagination to capitalize on the ideas as it was to get them.

SELECTED REFERENCES

1. Alger, J. R. M. & Hays, C. V. *Creative Synthesis in Design.* Prentice-Hall, 1964.
2. Allen, M. S. *Morphological Creativity.* Prentice-Hall, 1962.
3. Buhl, H. R. *Creative Engineering Design.* Iowa State University Press, 1960.
4. Gordon, W. J. J. *Synectics.* Harper & Row, 1961.
5. Haefele, J. W. *Creativity and Innovation.* Reinhold, 1962.
6. Osborn, A. F. *Applied Imagination.* Scribner's, 1963.
7. Pleuthner, W. "Brainstorming." *Machine Design,* Jan. 12, 1956.
8. Taylor, C. W. *Widening Horizons in Creativity.* Wiley, 1964.
9. Von Fange, E. K. *Professional Creativity.* Prentice-Hall, 1959.

6

Design Attitude . . . You Have to Want To

It takes more than intelligence and training to get something done. Just because people are bright and have taken lots of courses doesn't insure that they can produce. Some psychologists have been quite concerned over the reasons why so many brilliant children fail to achieve greatness when they mature; even more puzzling, why many of these children wind up in vocations requiring only mediocre skills and intelligence. Obviously, horsepower isn't enough.

There are many psychological constraints of genealogical and environmental origins that interact to inhibit the attainment of our full intellectual capabilities. Many of these are deep-rooted, complex, and inaccessible to alteration, except perhaps in a limited way by psychiatric techniques. Fortunately, however, individuals have a great deal of conscious control over their intellectual capabilities. This self-control can be summed up in one package called *attitude.* What then is a good attitude for a design engineer?

IF YOU WANT TO–YOU WILL!

This is one of the fundamental attributes of a good design attitude. It represents tenacity and a determination to "rassle" the problem until somehow, someway you work out a method to solve it. If you don't really care whether you solve it or not, you will give up when it gets tough.

Hard work and determination Have never caused extermination.

Determination and **drive** are two admirable traits in any individual. A survey was conducted among a group of young men who had one remarkable thing in common–they all had become millionaires before they were 30 years old. It was noted that they all had two unique characteristics: an uncommonly strong desire to make money and a tremendous will to succeed.

The human potential for accomplishment is probably one of the greatest undeveloped resources on earth. No one knows the heights a person can attain. Occasionally, and accidentally, some unusual circumstance will motivate a person to do something he or she otherwise could never do. We have seen unusual physical accomplishments under severe conditions of panic or fear (a mother single-handedly lifting a car off her child). Under hypnosis, people can be called upon to do mental or physical feats they are incapable of performing when conscious.

Any good mind can be unusually ingenious when properly motivated. What makes a person want to do something? What is the source of motivation? There are several basic drives: ego (i.e., pride, desire for praise, recognition, prestige), survival (hunger, fear), and sex. Of these three, ego is the noblest and most complex driving force.

Nearly everything you try to accomplish beyond satisfying your animal needs is motivated in some way by your ego. Ego is the voltage that makes one's motor run–the higher the voltage the more action there is. A person's ego-strength varies considerably with his or her general mental health. On some days you feel like you can take on anything and anybody without effort and on others you could sell out for a nickel and make a profit.

Success and achievement are largely measured by the type of "pay-off" one needs in order to realize a sense of satisfaction.

PORTRAIT OF A TYPICAL ENGINEER

What is a typical engineer? What kind of person becomes one? Raudsepp (4) correlated the results of a large group of studies made by psychologists to determine the personality, habits, attitudes, and abilities of the engineer. Here is what the "typical" engineer looks like:

Personally:

Reserved, self-sufficient, independent, well adjusted.
Has little interest in people—is object and idea oriented.
Tends toward shyness and is not emotional or impulsive.
Industrious, determined; serious, sincere, honest, orderly.

Intellectually:

Very intelligent—in the upper ten percent of the population—prefers math and science.
Tends to be narrow; has few cultural interests; reads little outside of field.
High mechanical and visual ability but is poor at communication.

Socially:

Not gregarious; has a casual relationship with colleagues.
Superficially friendly; avoids leadership in group activities.
Insensitive to needs of others, has few intimate friends.
Tends to conform socially, dislikes chit-chat.

At work:

Action oriented, hard worker, eager beaver, well organized.
Highly motivated to achieve a successful solution.
Prefers concrete, orderly mechanical tasks; tends to be conventional.
Works hard to avoid criticism and failure; cautious, and conservative.
Respects authority, but dislikes being supervised.

At home:

Excellent spouse and parent—family oriented.
Solid middle-class suburbanite; willing worker in community projects; interested in gardening, home repair, and crafts.

At play:

Active hobbyist; primarily mechanical hobbies; strong interest in outdoor life.
Spectator of team sports; active participant in noncompetitive sport activities. Willing participant in family-oriented recreation.

To sum up: An intelligent, hardworking, independent, action-oriented, solid citizen and family person.

A high ego-strength usually sets high goals—a low ego-strength requires only mediocre achievements. A measure of the quality of a person is the goal chosen to be accomplished.

You will not voluntarily undertake any task with enthusiasm unless you can anticipate a pay-off—some kind of personal satisfaction. These pay-offs can run the gamut from selfishness to idealism. Each of us has a list of high priority motives that stimulate us to full throttle activity. Money, praise, and recognition are high on most lists. Fear (of failure or ridicule) and hate (for a competitor, or a way of life) are frequently powerful motives. Some people are challenged by the accomplishments of a great person in their field, and hero-worship is their motivating force. Others are highly motivated by the nonhero (the mediocre performance of an associate) and work hard to excel where others fail. People to be pitied are those who are so neutral that they have no inner motives except to win a pat on the head for doing what they were told.

The world sorely needs self-motivated people who "want to"—and for noble reasons. Engineering design activity requires engineers who are highly motivated and self-starting and who have a strong will to succeed and the ego-strength to remain confident that ultimately they will get the job done and done right—no matter how tough it is.

YOU HAVE TO WORK AT IT

"Wanting to" provides the motivating force in design activity—but it takes stamina to get the job done. Just to have voltage on the motor doesn't help if the windings overheat and the motor fails under the work load.

Thomas Edison once remarked that invention is 1 percent inspiration and 99 percent perspiration. There is a great deal of agony, anguish, worry, frustration, hard work, and failure in the accomplishment of an engineering design. It takes a lot of persistence and determination to beat back nature and have things line up the way you need them.

In this age of specialization there is a strong tendency to adapt the adage:

> If at first you don't succeed, try again.
> If that fails, say "to hell with it"
> And turn it over to a specialist!

One of the most admirable traits one can adopt in engineering design is patient determination. "Trying again" involves an amount of risk and gamble and a fairly philosophical attitude about failure. Design activity requires a high tolerance of failure—an open-minded attitude that every failure produces useful information. A design engineer has to accept the fact that experimentation generates failures, and failures generate experience.

> He who pales
> When it fails
> Will never rise
> To capitalize.

Every idea that fizzles—every design that fails—produces useful information—the prime ingredient for a revised and modified alternative. The development of an objective attitude toward

A sure way to mediocrity is to quit too soon.

failures is a necessary attribute of a good designer. If every failure became a personal disaster, a design engineer would be a mental case within a week.

> Pity the emotional engineer,
> Whose life is filled with constant fear
> That every fault in his design
> Will wipe him out before his time.

One sure way to mediocrity is to quit too soon. Creative design activity, the engineering problem-solving situation, is loaded with frustrations. Ideas don't come easily; every design engineer sometimes is moved to say "to hell with it" in total frustration and discouragement. It is a normal and typical phase of the creative process. A considerable amount of patience and confidence in yourself is required to capitalize on frustration.

> He who throws in the towel too soon—
> Will likely be wet behind the ears!

The stress of frustration is a necessary stimulation to incubated ideation and the mind needs time and conscious diversion to work on it. It is a healthy design attitude to be able to accept discouragement and frustration as a necessary and useful part of the process —to welcome the opportunity to drop the project for awhile until your ego strength revives—and to have confidence that when you get over it, you will have a fresh insight into the problem. There is always another way to do it—you just have to work at it.

Success depends heavily on quitting when you are "stopped cold" and then coming back later to have another go at it—even when you haven't the slightest idea, at the moment, what to do next. The "*if*" in "if at first you don't succeed . . ." is a certainty in engineering design. Doing it over more than once is standard procedure, since design is an iterative process. You have to learn to accept each try as a necessary step toward success.

> If at first you *do* succeed,
> And move to praise this noble deed,
> Stop and look—for what you cheer
> May have been caused by something queer!

IF WE MADE IT—IT CAN BE IMPROVED

Anybody can find fault with anything any time. It takes no talent at all. As a matter of fact, we're all quite adept at it. There is nothing more irritating than to have someone continually run something down and yet never have one good suggestion for improving it.

It's one thing to condemn and find fault. It's an entirely different ball of wax to evaluate a situation in order to improve it.

This is positive dissatisfaction, a very necessary attribute of a design engineer. A good designer is much less concerned with what is wrong than with what is needed to make a design better.

Every design, product, or process conceived can be improved. A perfectly good design can be rendered obsolete by a mere spoken word—the germ of a new idea. Nothing that now exists can be defended against change. In a dynamic society, the environmental constraints that forced the existing design do not remain constant. The reasons why it was done "that way" continue to change. Thus, the design must be changed to improve effectiveness.

YOU HAVE TO CHANGE IT TO IMPROVE IT

One of the greatest threats to the effectiveness of a design engineer is his or her own natural, almost instinctive, resistance to change. Any change is a threat to our comfort, security, status, or pride. It's part of our natural defense mechanism. We are creatures of habit—we couldn't function as efficiently as we do if we couldn't learn to do routine tasks without conscious thought. Consequently, any change that encroaches upon an established habit pattern is a threat to our ability to function without hardship or effort.

> **GO AHEAD AND CHANGE IT!**
> You may be surprised at what will happen.

Resistance to change is one of the greatest problems that any product innovation has to overcome. It is one of the reasons why millions of dollars are spent on advertising, promotion, demonstrations, public relations, etc. It is the reason why engineers must learn to communicate (sell) their design ideas to their supervisors and their company. It is even a constant threat to an engineer's acceptance of modifications in her or his own design.

It is very difficult for a design engineer to accept a suggestion for a design change that would virtually wipe out days of work—even when it is obvious that it would be better. It is not uncommon for some design engineers to work hard to defend and make-do an idea that is in difficulty, rather than throw the whole mess out and start over.

When an engineer loses objectivity and measures success in terms of professional status, any shortcoming in the design becomes a personal threat. A design engineer's creative effectiveness is totally destroyed if criticism, suggestions, complaints, and failures begin to loom as threats to his or her ability as a designer. An exciting challenge turns into a living hell.

It is to the credit of design engineers when they can accept the fact that any idea they have can be improved upon; that complaints and criticism can point out valuable improvements; that every failure opens up an opportunity for another try for an even better approach; that the more alternatives they can consider the better the possibility for a truly new innovation—and, most important of all, that determination and hard work have never yet failed to pay off.

IT PAYS TO GO FIRST CLASS

It pays to go first class—this is a well-worn cliché, but it's a design attitude that pays off in a quality product and builds design integrity. It takes very little talent to design a mediocre product. Engineers who are willing to go "cheap," compromise quality for quick profit, do only what will get by, take advantage of a user's tolerance, and operate through "loop-holes" are a dime a dozen. It takes first-class engineering effort to attain quality. And it takes a person with a strong sense of perfection to accomplish a first-class engineering effort.

FIRSTCLASSMANSHIP

The Ten Commandments of Design

1. Thou shalt have no other goal before quality.
2. Thou shalt not make any false images of thy product.
3. Thou shalt not bow down to the pressures of compromise.
4. Thou shalt not take the name of thy product to shame.
5. Honor thy customer and thy profit margin.
6. Thou shalt not kill thy sense of perfection.
7. Thou shalt not commit abuse of thy product's function.
8. Thou shalt not steal longevity from thy design.
9. Thou shalt not bear false claims against thy design.
10. Thou shalt not covet thy neighbors' designs, nor their design engineers, nor their production techniques.

A design engineer in the process of developing a design is faced with a multitude of alternatives, each demanding a decision and a choice. Every alternative has to be weighed against all the restrictions and constraints involved in the over-all project. There is always the pressure to reduce cost—to make an economical product—and there is a constant temptation to compromise quality in favor of economy. For every "give" there is a "take," and the final design is a string of compromises.

Quality in a product is measured not by price but by function. It is careful attention to purpose, during the designing, that creates value. More times than not, it is seemingly inconsequential and minute detail considerations that give the product stature and respect. As in most every other endeavor, "it's the little things that count." And, usually the cost of adding the little things is negligible.

PORTRAIT OF A DESIGN ENGINEER

What contrasts the design engineer from other engineers? What are the attributes of an engineer that make her or him a good design engineer? Raudsepp (3) also surveyed the attributes of the creative design engineer. Here is what the typical design engineer looks like:

Personally:

Very self-confident; willing to take calculated risks; open-minded to experience; highly tolerant of criticism; distrusts routine; dislikes regimentation; has strong initiative; uninhibited in communication of ideas; is a constructive nonconformist; imaginative.

Intellectually:

Strong background in fundamentals, with broad interests; very curious and observant; highly creative; persistent and patient in pursuit of a solution; looks for alternatives and ambiguities; very flexible; tolerant of ideas and innovations; has strong interests in mechanical, scientific, artistic and literary areas.

To Sum Up:

In contrast to other engineers, they take the creative nonconventional approach, are willing to take risks, are not afraid of failure, are idea-oriented and creative, and have a much broader intellectual scope.

A design engineer with a "first-class" attitude will insist on doing a job in the best possible way. The best way doesn't need to be the costliest way—there is a marked difference between "inexpensive" and "cheap." It is an attitude that prompts a continuing search for perfection—the optimum combination of parameters.

It is concern for reliability, safety, simplicity, function, economy, and attractiveness; an attitude that "if it's worth doing—it's worth doing well."

ALL IT TAKES IS TIME . . . (AND MONEY)

Design is a conceptual activity. It involves the creation of something that does not now exist. Thus, the design engineers are involved in the development of something beyond their own experience. They are constantly facing a void—and are being called upon to do something that has never been done before.

To work in such an environment requires a considerable amount of optimism and self-confidence. If you aren't convinced it can be done, you will stand an excellent chance of not doing it. An engineering design class was asked to investigate a seemingly impossible idea and determine if it would be possible to actually build and operate such a system. Approximately one-third of the group threw in the towel in a few days, convinced it couldn't be done. They were embarrassed by their more confident colleagues, who worked at it until they had found a way to do it.

Self-confidence and optimism are not necessarily related to technical competence or a vast knowledge of the subject area. Rather, these qualities are more related to having a strong faith in the results of conscientious hard work, the power of positive thinking, and the fertility of the imagination.

BE PATIENT!
When you're frustrated
Don't stomp and stammer.
Could be that your brain
Has an inept programmer!

The very fact that you have the level of intelligence that it takes to succeed in engineering is an indication that you have a mind with exceptional capability. That alone is sufficient reason to have a tremendous amount of confidence and optimism that you can succeed in any design situation.

It just takes time. After 50,000 years, we have yet to learn how to really take advantage of these marvelous brains we have. It would be wonderful if we could attach electrodes to our heads and program these fantastic computers just as we do our electronic computers. Instead all the inputs have to be filtered through a

> **DO SOMETHING! DO IT NOW!**
> **Or someone else will pull your plow.**

We have yet to learn how to really take advantage of these marvelous brains we have.

maze of emotion, prejudice, fear, and laziness. It is a wonder we ever accomplish anything. But by brute force and sheer determination, we can ultimately force ourselves to generate a very acceptable new solution to a design situation. You have to be patient with yourself.

SELECTED REFERENCES

1. Danielson, L. E. *Characteristics of Engineers and Scientists.* University of Michigan, 1960.
2. Kettering, C. F. "How Can We Develop Inventors?" Reprinted in the appendix of E. K. Von Fange, *Professional Creativity.* Prentice-Hall, 1959, pp. 223-234.
3. Raudsepp, E. "The Creative Engineer." Part 1, "Personality Traits," *Machine Design,* May 28, 1959, pp. 22-26. Part 2, "Intellectual Abilities," *Machine Design,* June 11, 1959, pp. 28-32. Part 3, "Mechanics of Creativity," *Machine Design,* June 25, 1959, pp. 27-30.
4. Raudsepp, E. "The Engineer—Paragon or Paradox?" Part 1, "His Personality," *Machine Design,* Dec. 10, 1959, pp. 24-28. Part 2, "His Intelligence and Abilities," *Machine Design,* Dec. 24, 1959, pp. 29-31. Part 3, "His Interests," *Machine Design,* Jan. 7, 1960, pp. 25-28.
5. Von Fange, E. K. Chapter 2, "The Challenge of Thinking." *Professional Creativity.* Prentice-Hall, 1959, pp. 15-27.

7

Value Engineering . . . For Goodness Sake

Everything an engineer does costs money, and lots of it. Every time an engineering project gets underway, someone is worrying about cost. Someone is also worrying about quality. When cost is the only concern, you are apt to wind up with a cheap and chintzy piece of junk. When quality is the only objective, you will quite likely produce a gem that will be quite expensive. But when cost and quality are both sought after, you are concerned about value—getting the most for your money—getting a *good* design.

There is an aspect to value engineering other than merely getting the most economical design. While economic value concerns the functional benefit in relation to cost to produce, a product has no value at all if no one needs it. When a device is in short supply, is vitally needed or highly desired, its value can far exceed any cost considerations. Actually, there are three significant dimensions to Value: Functional Value, Need Value, and Esteem Value. If a product has a big psychological value (personally important), a sociological value (the "in" thing), or symbolic value (success or status symbol) it will be valued independently of its cost. If a design has vital need by a large component of our society its quality and reliability and safety are concerns that exceed concern for cost. Even so, it is the design engineer's duty to assess the cost and seek the best compromise of cost to need.

Behold the valuable design engineer, Who values the value of value quite dear!

FUNCTIONAL VALUE

Value engineering is a new facet of engineering design activity that is concerned with cost reduction through design or redesign. The field evolved primarily out of the defense industry as an effort to eliminate needless and costly design and production

procedures. It was discovered that many defense-contracted items were far more costly than their commercially available equivalents. Many of these higher costs were traced to rigidly-enforced detailed standards and specifications that were written into contracts to reflect anticipated needs and requirements. In some cases these arbitrarily severe and overly conservative specifications were set to assure high reliability and to anticipate a broad range of conditions and requirements that, in actuality, have a very low probability of occurrence. As a result, large quantities of items were produced which far exceeded the capabilities and quality actually needed.

Cost-reduction procedures in manufacturing are not new. The entire field of industrial engineering is devoted to efficient management of the production process for the express purpose of reducing the unit cost of the product. This reduction is an absolute necessity in mass production. However, while industrial engineering is concerned with manufacturing efficiency and plant management, value engineering is concerned with the development of a design that attains its optimum functional potential in the simplest, most direct configuration that will inherently reduce the over-all cost of the product. It is economy by design in addition to economy by manufacture. It is a design effort to reduce the cause of cost by economical production techniques, rather than an effort to reduce the effect of costly design decisions.

There is, however, a built-in dilemma when one sets out to reduce the causes of costs in design. Immediately, there arise the conflicts of functional necessity, economic necessity, and some basic and philosophical attitudes concerning the practical definitions of value and quality.

THE COST OF QUALITY . . .

There is a basic conflict between reducing cost and maintaining quality. The motives are diametrically opposed, for cost reduction generally results from having "less of" and improved quality requires "more of."

Although we sing "the best things in life are free," it doesn't necessarily hold in engineering design. The best things in design are not free; they are often obtained at considerable cost. There are several factors that contribute to the cost of quality:

- It is ironic that many of the substances and materials which have superior qualities and attributes are either in scarce supply or are extremely difficult and expensive to obtain. Many of the alloying metals used to make high-strength, high-performance steels are scarce and expensive. The noble metals are also scarce and highly valued.

- Nature demands a healthy price for its secrets, and many of the materials and processes that contribute to quality products are quite costly, demanding large expenditures in energy, labor, and mechanical processing equipment.
- Demand vs. supply contributes to cost when the volume demanded has never justified the development of more economical methods of production. Many operations contributing to high quality are still accomplished by highly-trained skilled craftsmen on a piece-work "hand-made" basis! Such costly operations could be done at much lower unit cost by specially designed machines only if the volume of production would justify the development of the machines.
- It has become generally accepted in our society that quality is expected to cost more. We associate quality with high cost, and expect something to be better if we pay more for it. Thus, the cost of better quality is actually propped up because the consumer is willing to accept a higher price for better performance. It is not uncommon for the more expensive products to have a higher profit margin because of the status of quality.
- Quality is also measured by reliability and elegance. This creates the demand and necessity for "extras" in the design. Here the expense of quality is derived from expenditures for back-up systems, extra strength, triple plating, safety features, fail-safe devices, precision matching of parts, automatic controls, self-monitoring maintenance devices, noise reducers, vibration isolators, etc., etc.
- Sheer luxury and excess are often (mistakenly) used to measure quality. The inclusion of expensive decorations or finishes, supplemental devices, and gadgets adds nothing to the performance but contributes a measure of status to the owner.

Thus, there are many factors that tend to cause "quality" to cost more, although many of these are synthetic and irrelevant.

> Quality is never cheap—
> But needn't have a price that's steep.

The ultimate purpose of good, effective, value engineering is to obtain true quality at a reasonably low cost.

. . . expensive decorations or finishes, supplemental devices and gadgets add nothing to the performance but contribute a measure of status to the owner.

WHAT MAKES AN ECONOMICAL DESIGN?

THE CRITERIA OF GOODNESS
What Makes a Good Design Good

• Low Cost	• Ruggedness
• Simplicity	• Functionality
• Reliability	• Attractiveness
• Usefulness	• Desirability

An economical design is an optimized design—getting the most for the least—it is a design without excess—a design with every part exactly sized and shaped for its intended purpose. No material is used that isn't needed. The material is placed and shaped to function at the maximum allowable stress level throughout. Weight, number of parts, number of assemblies, amount of machining, complexity of shape are all minimized. The least expensive (but most appropriate) materials are selected and fabricated in the least expensive manner with the least amount of manual labor. In general, economy in design is the development of a product to do only and exactly what it was designed to do, and no more, with the least expenditure of materials, resources, production effort, and labor.

WHAT MAKES QUALITY?

Quality in engineering design is defined by such terms as:

- Sturdy
- Reliable
- Long life
- Low maintenance
- Trouble-free
- Lasting appearance
- High accuracy
- Excess capacity
- Low noise level

High quality connotes excess—it connotes exceeding the minimum requirements; it also connotes anticipating the unusual and infrequent demand. Quality means long life, high reliability, and low maintenance. It means extras—extra-strength materials (nickel steel rather than cast iron); extra components (back-up systems); more precise machining; more adjustability (instrumentation and controls); highly durable finishes (chromium plating rather than paint); etc.

WHAT MAKES IT "CHEAP"?

Economy can be carried too far and, when it is, a "cheap" product results. The typical product resulting from a "cheap" design effort is unreliable, faulty, and fails to live up to its purpose.

> Seek and ye shall find
> That squeaks lead to a bind,
> And squawks precede the balk
> That wipes out what's designed.

"Cheap" and "chintzy" product design has many characteristics, such as:

- Flimsy
- Non-adjustable
- Non-repairable
- Cracks
- Warps
- Dents
- Rusts, peels
- Corrodes, chips
- Overheats
- Rattles
- Noisy
- Inaccurate
- Unreliable
- Drifts
- Wears out

This is to say that we identify a poor design as one that fails to meet its design function, except on a marginal basis, and soon fails because variabilities in its use exceed its capability.

THE VALUE DILEMMA

"You can't have your cake and eat it too!" "You get what you pay for!" "You have to pay extra for quality!" "There's no point in paying for something you don't need!" Sound familiar? Then what is value?

- Is it paying more to go first class?
- Is it getting the most for your money?
- Is it getting only what you need?
- Is it the lowest-priced product?
- Is it the highest-priced product?
- Is it the most reliable product?
- Is it the most attractive product?
- Is it the long-life product?
- Is it the product that won't go out of style?

And further:

- Does "cheap" mean economical?
- Does quality mean excess and lavishness?
- Does quality mean overdesigned to anticipate every conceivable circumstance?
- Does economy necessitate limited function and design life?

Value is relative. The criteria upon which it is defined are varied and complex. Ideally, the best value is the best performance for the least cost. But how much is good performance worth when weighed against initial cost? Can one assume that good performance and long life will override the higher initial cost of quality design?

The value of a product is its worth to the user. It is the highest quality that is *needed* for the *function* it will be *expected* to perform, at the lowest possible cost. The dilemma that arises for the design engineer is how much quality is really needed to represent value to the customer for a given cost. That is, the designer finds herself or himself in a position of judging what represents value to the customer. Since quality is usually gained at a sacrifice of cost, quality is negotiable in assessing value criteria.

HOW LONG SHOULD IT BE VALUABLE?

One of the most expensive attributes of quality in product design is length of trouble-free service—product life—long-term reliability. It is also the most controversial and troublesome criterion in value assessment. There is merit in designing all products for an indefinite maintenance-free service life. However, there are many strong and practical arguments *against* achieving the "perfect" design.

Now, here's a feat
That's hard to beat:
A plan so neat
And so compleat,
They'll never bleat
"It's obsolete!"

Long-life design with high reliability is sometimes the prime criterion of a design effort. The gear-reduction units of large ships must be designed to give continuous full-load service for the life of the ship (exceeding 20 years) without failure or repair. This is necessitated by the prohibitive cost of replacement and loss of revenue should a repair have to be made. Obviously, the cost of indefinite life is easily justified in this case.

In mass-produced consumer products, long-life design is the subject of a controversy centered on the issue of product obsolescence. It has been argued that "life-time" products would depress a market and reduce production capacity, for they depend on continued sales, product turnover, and replacement. Some people claim that the public doesn't demand quality and prefers to replace a product frequently in order to "keep up."

This dilemma concerning how "good" a product has to be has extended to a concern for who or what is responsible for the product's obsolescence. The design engineer is faced with the charge of over-designing as well as under-designing. In an effort to establish a product's life expectancy, the designer is subject to the charge of causing a product to become obsolete and forcing its replacement.

There are three types of obsolescence according to Raudsepp and the editors of *Machine Design* (15):

- **Obsolescence of function**—caused by technological progress, the creation of new and better products.
- **Obsolescence of quality**—caused by built-in design limitations which limit product life.
- **Obsolescence of popularity**—caused by customer desire for new and up-to-date products.

There is no quarrel with the old being replaced by the new due to technological innovations: the horse and buggy by automobiles; pony express by telegraph; wringer washer by automatic washer; steam locomotive by diesel; etc. This is progress, and everyone stands to gain from the evolution.

However, the "newness" cult that has been nurtured in our society by the advertising media has its own controversial overtones. The whims, fads, prejudices, and preferences of the consumer have been discovered to be manipulatable and lucrative. Personal satisfaction is closely coupled to the status symbols of a materialistic society. Keeping up with the Joneses, being the first in the neighborhood, having something new, and keeping up to date are motivations and desires that sales organizations play upon to increase the flow of their products. Faddism has had a profound influence upon styling and appearance in product design and is also reflected in a demand for frequent and expensive model changes.

Design engineers are urged to repackage their basic products continually—devise gimmicks and accessories; create a variety of model choices. These variations in style and product appeal, over and above the basic product function, place an additional burden on the design effort. Frequent styling changes and a large number of model variations of a given basic product line necessitate increased production costs. Every styling change requires an extensive retooling effort; each model variation requires additional manufacturing stages.

In a highly competitive market, most of these additional costs must be absorbed rather than passed on to the consumer. Thus, there develops the vital concern for cost reduction. And, the design engineer is called upon to consider seriously the disadvantages of developing a product with the level of quality that would cause the product to outlast the customer's desire to own it. After all, what's the point of building a product that will last 10 years when the customer is going to replace it with a newer looking one every three years? Think how much money you could save if it were designed to last only three years.

> He who skimps
> While trying to save
> Digs his product
> An early grave.

THE FINITE-LIFE ARGUMENT . . .

The issue has not been resolved. There are logical arguments on both sides. The editors of *Machine Design* magazine, in a recent report (15), surveyed 500 design engineers and 500 engineering supervisors about the ethics of planned obsolescence of product quality—finite-life design. Here are some of the arguments used on both sides of the question:

In favor of finite-life design:

- It's necessary. Our economy demands a high production rate. Product turnover makes more jobs, makes more products available to more people at lower cost.
- It's foolish to build a product to last longer than it's needed.
- No one could afford an infinite-life product.
- It's a waste to build more life into a product that will eventually be scrapped anyway due to technical progress.
- Our accelerated technological advancements are creating new developments every day. No company can afford to bet its future on high quality products that are out of date.

- Short-life products encourage continued design development and innovation.
- The public wants an excuse to purchase new models.

Against finite-life design:

- It's unethical, immoral, and disgraceful.
- It cheapens our product image—develops a national reputation for shoddy merchandise, and will encourage foreign competition.
- It wastes our resources—generates junk.
- Basically, a consumer wants good quality and honest workmanship.
- It discourages research and development of new and better products.
- It discourages good design work and depreciates the conscientious efforts of a design engineer.
- It tends to lower the standard of living by creating cheaper, low-quality goods.
- It places an unnecessary economic burden on the consumer for continued replacement and repair.

Obviously, finite-life or planned-obsolescence designing is not a clear-cut issue. In general, it appears that engineering management regards it as a healthy design practice necessary to a dynamic economy. They argue that the customer is the controlling factor and that poorly designed and poorly manufactured products will not succeed anyway.

Consider now a punk design
That wore out long before its time,
The "make-a-buck" priority
Eliminated quality.

Fundamentally, no one can argue against good design. The design engineers should always develop a product that satisfies its intended purpose and gives good trouble-free performance as long as the customer needs it. There is no excuse for shabby design work, nor can short-changing the product's usefulness be condoned.

PHILOSOPHICALLY . . .

Value in engineering design is a debatable issue. It can be interpreted in many ways. Fundamentally, design engineers must somehow achieve their purpose in the most economical way. At the conceptual stage of the design process, it is impossible to estimate the actual cost of the alternatives being considered. It is sometimes not even possible to compare the relative costs of alternative plans. However, economy is derived from some fundamental attributes which can be observed to insure the likelihood of over-all cost reduction.

The following are some of the important cost-reducing considerations that should be observed at each phase of the decision-making process:

- **Eliminate!** Is the function actually needed? Stop for a moment to consider if you can eliminate this phase of the design. There is nothing that will cut costs as dramatically as finding a way to eliminate a part or assembly altogether. This is the first criterion of value engineering.

> If cost reduction's your desire,
> Then here's a way guaranteed sure-fire,
> Give your design a hard review,
> And reduce the parts from ten to two!

- **Simplify!** What is the simplest possible way? The second fundamental rule in value design is simplicity—get the least number of parts—with the simplest shapes.

> Simple Simon made his pile
> With simple parts of simple style.

- **Duplicate!** Strive for symmetry. Develop the design to contain as many identical parts as possible. It concerns a choice of making one tool for ten parts or ten tools for ten parts.
- **Standardize!** Choose a configuration that will use off-the-shelf, high-volume production components. Don't custom design when you can buy it ready-made!

- **Slow!** Watch for high relative velocities; whenever possible, seek to reduce relative motion between parts. High velocities cause high stresses, increase the wear rate, generate heat, and increase noise, all of which necessitate additional design compensation.
- **Smooth!** Cyclic variations in velocity create accelerations and dynamic loads which create a need for balancing, higher-strength parts, more sophisticated bearings, vibration-damping devices, and special design considerations to combat fatigue, fretting, and noise.
- **Cool!** Avoid extreme operating temperatures which increase corrosion rates, reduce strength, deteriorate materials, and necessitate the selection of more expensive materials and the addition of supplemental design considerations to reduce the heat.

> He who can't eliminate
> All that useless excess weight
> Will surely find the battle lost
> When he attempts to cut the cost!

- **Lighten!** Weight is paid for by the pound. Excessive weight is costly in material costs, static and dynamic loading, oversized parts, increased costs in machining, handling, and shipping.

These criteria dictate inherent cost savings throughout the design process. They are the fundamental guidelines for optimizing the design at the conceptual stage.

HOW FAR CAN YOU GO?

There is, seemingly, an endless opportunity to reduce the cost of a product. The task of cost reducing is similar to the task of ideating—the more you work at it, the more successful you are. Eventually, however, you have to cut it off and move it out the door. And then, invariably, you will come upon a new approach that is better than the one you have completed.

Great savings are achieved by redesign of the existing product. Experience with the existing product is an excellent guide to new approaches in cost savings. Most companies have found that it pays handsomely to devote a considerable amount of engineering effort to product redesign and the development of more efficient manufacturing techniques. Many valuable design changes are prompted by savings they will generate in the manufacturing process.

"A bird in the hand is worth two in the bush." It is considerably easier to develop a less costly solution when the current model is available and all the cost facts concerning its production are known. Also, specific alternatives for cost savings can be actually compared. Thus the application of value engineering techniques at this stage is a specific problem-solving process that generates specific answers. Whereas, in the development stages of design, the economic value of a decision is only a generalization.

Although many companies have continuously been able to reap a harvest through value engineering, there is a limit to how far one can go. Most cost relationships are not linear. They tend to follow what could be called "Harrisberger's Hyperbolic Law": *The more you work at it—the less you accomplish.* Or, according to "Pareto's Law" (9): *80 percent of the results come with the first 20 percent of time and effort.* The first revision reaps a substantial savings, but the next has less, and so on, until the cost to change it isn't worth the savings. There are many examples in product design where the cost of refinement far outweighs the benefits. And in a similar way there are frequent traps that can occur where a design modification intended to cut costs actually required a more costly technique of manufacture. But there are also situations where a less costly low-strength material could be replaced by a more costly high-strength material for an overall cost savings due to reduction in weight and size.

Effective cost reduction in the design stage requires a thorough analysis of the overall effects of a seemingly logical cost-saving proposition. Studies must be made to determine optimum savings—the compromise that achieves the lowest overall unit cost. You can go too far and end up increasing the cost.

HENCE, THUS, AND THEREFORE . . .

From an engineering design point of view, quality is reliable trouble-free performance without excess. It is a clean, lean design that performs smoothly, quietly, with minimum maintenance.

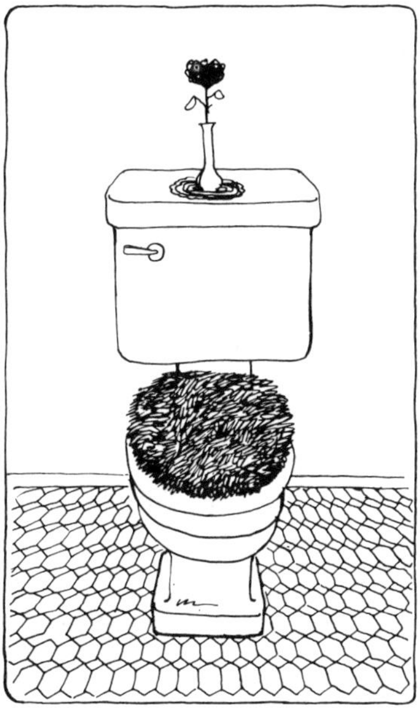

It is a clean, lean design that performs smoothly, quietly, with minimum maintenance.

Quality is engineered to meet functional requirements—tested and proven in service for reliability under all the conditions of use in which it is expected to serve. It is the direct result of optimizing a design for function. It has no luxury, no features that have marginal utility, yet it is not crudely shaped nor commercially unattractive.

Although there are a variety of conditions that create the customer's attitude toward value, the design engineer is morally obligated to develop a product of the highest quality and lowest cost. This implies then that the design should carefully provide for the range of usefulness the customer will demand and provide the best solution to meet these needs at the lowest possible price.

The planned-obsolescence controversy arose because products have been made at a sacrifice of quality, so that the product failed to live up to the needs of the customer. Good value engineering can increase the quality of products without destroying the product's competitive price.

WHEN MONEY IS NO OBJECT

In the vast majority of engineering design activities, getting "the biggest bang for the buck" is the overriding concern. The design engineer is constantly striving to achieve the best ratio of benefit to cost. But there are many situations wherein function, safety, and reliability override any concern for cost.

The customer is willing to pay whatever it takes when functional safety or reliability are at stake. This has been dramatized most recently in the NASA Space Program, the commercial aircraft industry, the nuclear power industry, and the current emphasis on environmental protection. In the cases of nuclear power plants and commercial aircraft, there is an unprecedented functional need coupled with a massive concern for the safety of large numbers of people. Because cost is secondary to function, there is a demand to employ the most sophisticated state-of-the-art concepts and techniques. The design engineer is mandated to go "all out" in research, creative exploitation of alternatives, experimentation and testing to achieve the desired outcome. Exotic materials, redundant back-up systems, computer managed controls, and human factors technology are all routine components of the design. Thus, the importance of function and mass need creates a demand that puts high value on sophisticated engineering solutions without concern for cost.

Another area where cost is secondary is the super project, such as the NASA Moon and Space Shuttle Programs, supersonic aircraft, the Aswan Dam, and so on. In these engineering projects the overall cost is so massive that one cannot afford to have the project fail because a part of the project was not given the fullest attention. Thus, the design engineer on each part of the project is encouraged by the *need to do everything possible* to produce a perfectly functioning component or system.

THE VALUE OF APPEAL

There is an element of customer/client psychology that cannot be overlooked by the design engineer sensitive to the value of the design. The customer/client will attach a personal value to a product or design derived from a personal assessment of its importance. The value derived from this importance transcribes in a willingness to pay more for "extras" in add-on functions, superior materials and finishes, luxury gadgets, and additional options and conveniences—all of which contribute nothing to the basic function of the design. These appeal values have been described by S. F. Love (9) as esteem values derived from psychological, sociological, or symbolic needs.

The importance of appeal/esteem values is particularly great when the product is being designed for a mass market—for example, appliances and automobiles. The marketing people are obviously well aware of these powerful customer motivators, as can be seen by the appeals made in advertising. These customer motivators reflect back to the design of the product in the form of demands for options and variations to the basic function of the product and a willingness to pay a disproportionately larger price for it. It is the appeal value that urges design engineers to "let themselves go" and "live it up" with all kinds of creative and luxurious alternatives.

THE VALUE OF ESTEEM

Psychological value: The satisfaction of owning the best you can get, the "top of the line"; superb engineering, quality of materials, and craftsmanship.

Sociological value: The satisfaction of owning something that benefits and pleases associates, friends, and community.

Symbolic value: The Rolls-Royce/Cadillac syndrome—ownership that sends status messages of wealth, patriotism, vitality, intellectual sophistication, etc.

SELECTED REFERENCES

1. Allen, C. J. "Value Engineering—Weighing the Alternatives." *Consulting Engineering*, Nov. 1974.
2. Eades, B. W. "Value Engineering as a Factor in Product Design." *Electrical Manufacturing*, Nov. 1959.

3. Frick, H. "Design Guide to Value–II." *Product Engineering,* Mar. 2, 1964.
4. Heller, E. D. "Cost as a Design Parameter." *Journal of Value Engineering,* Sept. 1962.
5. Huggins, R. T., & Hall, F. "Second Lesson in Value Engineering." *Product Engineering,* Mar. 30, 1964.
6. Huggins, R. T., & Hall, F. "Third Lesson in Value Engineering." *Product Engineering,* June 22, 1964.
7. Leslie, H. L. G. "Design Guide to Value." *Product Engineering,* Oct. 28, 1963.
8. Leslie, H. L. G. "What Value Engineering Does for You." *Product Engineering,* Sept. 30, 1963.
9. Love, S. F. "Value in Design Takes Many Forms." *Design Engineering,* Dec. 1976.
10. Maidanchik, B. I. "Value Analysis of Value Engineering Design." *Russ Engineering Journal,* Sept. 1974.
11. Orr, W. A. "Value Engineering during Design." *Proceedings, Society of American Value Engineers,* 1972.
12. Pugh, S. "Quality Assurance and Design: The Problem of Cost versus Quality." *Quality Assurance,* March 1978.
13. Ruggles, W. F. "First Lesson in Value Engineering." *Product Engineering,* Dec. 23, 1963.
14. Tangerman, E. J. "Value . . . The Emerging Emphasis in Design." *Product Engineering,* May 15, 1961.
15. "The Ethics of Planned Obsolescence." *Machine Design,* Aug. 27, 1964.

8

Style . . . Making It Look Good

Who cares how it looks as long as it works! After all, function is what really counts in engineering! This is the engineer's traditional point of view toward design responsibility. But, the situation is changing. More engineering products are being utilized by more people. We are fast approaching the day when our environment will become a monumental clutter of gadgets, appliances, and machines.

Everywhere we look we see engineering hardware. It dominates our visual environment. Everywhere we go—at home, on the streets, in the fields, in stores, offices, classrooms, in the air, on the water—there are products designed by engineers.

Obviously, if our entire lives are enmeshed in the hardware of our materialistic culture, it will have a considerable effect, not only on the physical appearance of our environment, but upon our lives as well. People are easily influenced by what they see and what they do. The "things" in our lives affect us and we become emotionally involved with them.

The shape and color of machines and products become important considerations in the design effort when the products must involve people in their operation and use. The ramifications of the importance of appearance in product design range from

After all, function is what really counts in engineering!

an aesthetic appreciation for beauty to the practical and serious aspects of utility and safety.

Because more and more people are using engineered products, more and more industries are competing for the market. Such competition forces increased attention toward market appeal, i.e., the psychology of consumer preference and consumer acceptance. This concern, therefore, is reflected back to the design engineer. Even though design engineers are not obligated to profess an artistic talent for styling, they are nevertheless obligated to acknowledge the constraints and considerations necessary to make the product acceptable in a social environment.

The design engineers should be cognizant of the attributes of styling that affect the value of a product to the consumer. They must be aware of the aesthetic importance of the influence of shape, the psychology of color, and the honesty of good styling. Design engineers must face an increasing responsibility for the shape our society will be in from now on.

AESTHETICS! IN ENGINEERING?

How in the world did aesthetics get mixed up with engineering design? It got into it because people have a curious characteristic of becoming emotionally associated with the products they own and use. When customers have a choice to make between products, they base their judgment on both functionality and aesthetic appeal. It looks good to them because it satisfies their criteria for usefulness and their criteria for pleasing appearance.

We all exhibit a pride of ownership. We select a product because we like it—it gives us some satisfaction and pride in having chosen it. And something is pleasing because it appeals to our emotions. Pride of ownership is a reflection of our ability to like things as well as people.

We have all seen the extent to which some people will get "carried away" over their automobiles. They give them names (usually feminine); fuss over them; wash them; shine and polish them; buy doo-dads and gadgets to decorate them; and become highly indignant and defensive over any derogatory criticism of their performance. As a matter of fact, some people treat their automobiles much better than they do their families.

Marketing agencies are well aware of a tendency of people to become family oriented toward a product line, i.e., Ford owners tend to buy Fords again, Chevrolet owners remain loyal to the brand, etc. This tendency to remain loyal causes manufacturers to maintain a strong "family resemblance" from one model to the next. At the same time, people want a product to be up-to-date and modern. Thus manufacturers must make frequent changes in style and color even while they maintain family resemblance.

Any thing of beauty is emotionally pleasing. To be a satisfying pleasure to own, a product must look good; have a pleasing performance; and should fit into the environment in which it will be used. Thus, as our streets become filled with parked cars (and our fields become littered with worn-out cars), our cities become surrounded by factories, and our homes filled with machines and appliances; all, collectively, become a huge montage of man-made color, shape, and form.

The designers of all these products have a responsibility to maintain an environment that provides aesthetic satisfaction to those who must live within it. The scope of this responsibility includes a concern that the sensibilities and appreciation people have to beauty and aesthetic goodness are nurtured rather than smothered. Good architects recognize their responsibility for the visual impact that their buildings will have on the entire community. They strive to create a structure that enhances and contributes to everyone's appreciation of a beautiful environment. Design engineers must also take an aesthetic responsibility for the visual effect of their products. For, by abundance alone, the multiplicity of engineered products can dominate our environment just as a single large building dominates its neighborhood.

ENGINEERED CRAFTSMANSHIP

The evolution of technology has created its own aesthetic responsibility for our cultural heritage. The demand for products has necessitated the development of mass-production techniques. As a result, the work of craftsmen and artisans is being replaced by automatic machines and product development teams.

Prior to the era of mechanization, the design and construction of our useful devices was in the hands of highly skilled and talented people who devoted their lives to creating beautiful and lasting hand-made products. They were dedicated to a ministry of beauty through fine craftsmanship. Their work and their skill were highly valued. They were trained by years of apprenticeship to skilled masters. Their techniques were often jealously guarded family secrets that were divulged only to apprentices who showed the greatest promise of talent and skill. Many of the products of these highly skilled and talented craftsmen are preserved in our museums as priceless objects of art and craftsmanship.

Today, no craftsman can compete with the precision and skill that can be achieved by our modern production techniques. Hand-crafted products are no longer a match for mass-produced products, either in quality or cost. The finest hand-crafted product can be duplicated by the thousands, with superior quality, and at a greatly reduced unit cost.

APPEARANCE DESIGNERS

Since product engineering has replaced the skill of the craftsman, it also has replaced the artistry of the craftsman, and has inherited the responsibility for preserving and enhancing the artistic heritage of our mechanized culture. To meet this responsibility, the production industry now retains the services of a highly trained, extremely talented group of industrial product designers. These specialists in industrial styling are specially trained artists who deal with the aesthetics of three-dimensional form—applied sculpture. All of the automobile companies maintain a permanent and highly paid staff of stylists to design their products for appeal and good taste. There are many styling consulting firms who contract to "package" the products of a variety of industrial product lines.

> Style awhile,
> And brush away those sneers.

Although industrial stylists have a heavy responsibility to enhance the market appeal of a product (the only way their work can be directly justified), they nevertheless take a great professional pride in contributing to the cultural heritage of our age. Through the efforts of these specialists—not without the awareness and appreciation of the buying public—we are moving swiftly into an era that shows a refined concern for beauty and good taste in engineered products.

The stylist primarily "packages" the engineer's design, but the engineer is, nevertheless, responsible in many ways for the final shape. A woman can do wonders with clothes and makeup, but it still helps to have something to start with. So it is with the stylish product. Design engineers who appreciate the aesthetic value of their product's appearance will see to it that the stylist has something to work with that will achieve optimum appeal. They will also be sympathetic to the problems of the stylist. But above all, they will be conscious of the basic design attributes from which good form derives, and will evolve a far superior design because of it.

DON'T APOLOGIZE

"Well, it doesn't look like much but . . ."
"It looks kind of odd, but . . ."
"We haven't had time to clean it up, but . . ."
"It sort of looks like a . . ., but . . ."
"We still could do some more work on it, but . . ."
"I guess we could have made it look a little better, but . . ."

BE PROUD OF WHAT YOU DO!

- If it's worth doing, it's worth doing well, and it's worth taking some pride in how it will look to others!
- A good job needs no explanation!
- Good appearance is an asset, both to the product and the designer who created it!
- There is no excuse for not trying to make it look better!

JUST LOOK AT THE SHAPE IT'S IN

If your design prompts a remark like "Echhh!"—you have had it. If it looks that bad, you can bet that the rest of the design is no better. Aesthetic appeal is an inseparable combination of function and appearance. The functional simplicity that inherently enhances overall appearance is the hallmark of an optimum design. A good design will be clean and purposeful in shape. There will be no useless material, no needless and unsightly joints, holes, projections, etc.—no excess or redundancy.

The shape of all living things in our world is the result of functional evolution. Nature does not tolerate excess. Life evolves in response to the conditions and requirements of its environment. In nature, there is a functional purpose and reason for the shape and color of every living thing. It is only natural for us to relate form to function and to find a great deal of aesthetic appeal in those forms that epitomize optimum function. Students of form recognize the egg as an object of beauty, just as artists eulogize the beauty of the healthy, normal, human body as an example of optimum form and function.

The shape of modern aircraft is a good example of optimized functional design. The constraints that weight limitations and high speed impose on the design force an optimized configuration without excess. The result is a natural functional shape that is very pleasing to behold. It is an excellent example of functional design with virtually no additional styling—it achieves aesthetic appeal because it is functionally clean and has purpose of form.

Although a good design will basically have good form, it doesn't necessarily follow that an appearance designer is not needed. These people are experts on aesthetic appeal, and they are concerned that the product obey the laws of good composition, both in shape and color. They are also concerned that its function and its appearance are complementary and useful.

Skilled appearance designers will never compromise function to achieve appearance. Instead, they will strive to make sure that appearance improves the usefulness and value of the product. They also are sensitive to all the ingredients that improve the emotional appeal of the product. This sensitivity to human reactions includes an awareness of the psychological forces that are influenced by the color, shape, form, and texture of a product.

Thus, appearance designers can perform a useful service for design engineers as consultants in usefulness and good taste.

A CHECKLIST FOR GOOD FORM

Does it look smooth and clean?

Or:

- Do you have some protruding parts, sharp corners, ugly bolts and nuts showing?
- Do the things that show look like they were added on as an afterthought?
- Does it look like anyone with a little thought could have cleaned it up?
- Does it look fat, bulgy, ill-proportioned, sloppy, awkward, ungainly, heavy, flimsy, crude, rough, or make-shift?

Does it look functional and useful?

Or:

- Does it remind you of something else?
- Does it look like you could get hurt using it?
- Does it look like you would have to be careful not to tip it over, or bump it, or mishandle it?
- Does it look fragile, like it would easily get out of shape, or would rust, dent, get dirty; or is it rundown looking?
- Does it look like you could count on it to do the job dependably and not show wear and tear?
- Does it just look like something is wrong with it?

Then do something about it!

A perfectly functioning product may lack a great deal in usefulness unless careful attention is paid to the effect it will have on the user. The sensory effect it has on the user can range from physical hazard and psychological repulsion to great satisfaction and pride. The design engineer who fails to appreciate the implications of style and appearance and usefulness will not only fail to provide for them, but will see no need for seeking the valuable counsel of a good appearance designer.

HOW ABOUT MAKING IT LOOK FANCY?

Why do we make it look fancy? To encourage emotional involvement? To enhance its appearance and its market appeal? There are several good reasons for decorating:

Although a good design will basically have good form, it doesn't necessarily follow that an appearance designer is not needed.

- It increases emotional value because the modern consumer has been educated to associate decoration and embellishment with increased worth.
- It is a means for obtaining a periodic "new look" in order to identify newness.
- It identifies a new model that may have functional improvements that aren't obvious in its appearance.
- To create and maintain a "family resemblance" in the product line.
- To accentuate functional characteristics, and identify the product's purpose to the user.
- To cover up necessary joints and fasteners required for assembly or maintenance.
- To give the consumer an opportunity to choose a product appearance that will complement its surroundings.
- To eliminate the visual and psychological illusions created by reflections, poor visibility, and undesirable shades and colors which could create a hazard to the user.
- To provide protection of the product from wear and tear,

corrosion, dents, scratches, etc., so it will retain its appearance and appeal.

Like anything else, decorating can be too much of a good thing. It can be misused and carried too far. Decorating has been used to "gild the lily" and create artificial appeal for cheap and poorly designed products. Also, bad decorating is worse than no decorating at all. If it looks cheap, flashy, gaudy, or tiring, it creates consumer distrust and establishes a product image that is difficult if not impossible to live down.

> Glitter, clutter, gaudy gadget,
> How I wonder if you'll hack it,
> Out among the competition,
> Looking like an apparition!

To decorate a product effectively and tastefully requires a considerable amount of artistic skill. The appearance designer can be of great value at this point.There are few design engineers who have the artistic skill to make good styling decisions without the counsel of a good stylist.

Decorating should complement the quality of the product and enhance its function and good form. There are, however, many instances in product development where the appearance designer is retained to "package a Kludge." The pressure of competition and time schedules force hasty engineering design and development. In the interests of saving time and meeting competition, the products are "jury-rigged," using off-the-shelf components instead of refining the design into smooth, functionally efficient proportions. Decorating in this case becomes a cover-up operation to hide the mess. There is no more justification for decorating to hide a poorly designed product than there is to make claims that falsify its performance. It is just as unethical to misrepresent aesthetic value as it is to misrepresent design capability.

"AS LONG AS IT'S BLACK"

Henry Ford once remarked that you could buy a Model-T Ford in any color as long as it was black. Millions of black Model-T's were made before competition provided enough alternatives so that the public could express its preference. And when we are given the chance, we enjoy making a choice of color.

Color is a powerful medium of visual communication. It affects our morale—makes us happy or depressed. Color can make us sluggish and tired or peppy and exhilarated. It can be repulsive and sickening or nostalgic and beautiful. It can cause accidents; affect our accuracy in skills; create moods; influence our attitudes; affect our opinions; incite prejudices; encourage belligerence and rowdiness; make us feel warm or cold; and affect our ability to distinguish objects. It can make something look large, small, long, narrow, wide, or heavy. It can make it look hard, soft, old, new, faded, cheap, or expensive. It can create illusions that seem masculine, feminine, sporty, conservative, arty, etc. In other words, color has an influence upon just about everything we do that involves our feelings, attitudes, and psychological reactions. Color, like music, plays an important role in influencing our behavior and how we feel.

Everyone has a color preference. In general, the vast majority of men prefer blue, whereas women prefer red as their favorite color. However, color preference is a fickle facet of our emotional life; it varies with mood, season, time, prevalence, exposure, and status. We base many of our color preferences on good or bad experiences we have had that are somehow related to that color.

Color preference is influenced by lighting, hue, surface texture, surface area, reflectivity of the surface, etc. Our reaction to color is also affected by the amount of color we see, the combinations of colors that are present, and the pattern that is produced by color combinations. With all these variables to contend with, it is no wonder that considerable talent, skill, and training are needed to make good color selections that will affect others.

COLORSMANSHIP

To include color in the design of a product requires not only a need for good artistic skill, but it can create some difficult problems of manufacture as well. Good color design is based on two basic considerations: it must have a functional role, and it must have a strong aesthetic appeal.

SOME COLORFUL FACTS

Color preferences (in order):

Men: Blue, Red, Violet, Green, Orange, Yellow
Women: Red, Blue, Violet, Green, Orange, Yellow
Children: Orange, Red, Blue, Green, Violet, Yellow

Color visibility:

Most noticeable color on a white background (in order):
To men: Black, red, green
To women: Red, green, black

Maximum range of visibility:

Red – 3½ miles	Yellow– 1½ miles
Green – 3 miles	Blue – ¾ mile
White – 2½ miles	Violet – ¾ mile

Color combinations for distinctive legibility (in order):

Best	*Medium*	*Poor*
Black on Yellow	Black on White	Red on Yellow
Green on White	Yellow on Black	Green on Red
Red on White	White on Red	Red on Green

Maximum color brightness:

Bright light: yellow and orange
Dim light: yellow-green, green, blue

Effect on eye focus:

Far-sighted focus: Red, orange
No focus: Yellow, purple
Near-sighted focus: Blue, green

Color designations:

Tint – Pure color blended with white.
Shade – Pure color blended with black.
Tone – Pure color blended with gray (white and black).
Hue– The predominating color in a mixture.

Color can be used in a variety of ways to improve the product:

- **Operation.** In the human/machine relationship, color plays a significant role in reducing human error by increasing the visibility of critical parts, identifying hazardous areas, lowering eye strain, reducing mental fatigue, improving morale, etc.
- **Identification.** Color is used to identify grades of quality (limousines are usually black); to code function (in factories

electrical parts are blue, operating components are cream colored, danger areas are orange, etc.); to identify the manufacturer (Caterpillar tractors are yellow, Euclid equipment is chartreuse, John Deere equipment is green, etc.).

- **Function.** When heat is to be radiated or absorbed, the surface color can be a significant influence on the amount of heat transferred.
- **Visibility.** Products used in rescue missions are specifically colored to have long-range high visibility. High-speed vehicles (trains, aircraft, fire engines, etc.) use color for warning.
- **Appearance.** We have already discussed the psychological effect of color and the necessity for the aesthetic attributes of appearance. The use of color has a significant and almost overwhelming influence on market acceptance and sales appeal.
- **Cost.** The selection of color can have a considerable effect on the cost of manufacturing. Some difficult problems can arise when different materials, surface conditions, and techniques for coloring are to be combined into a color-matched assembly. An arbitrary color selection can create problems of obtaining a pigment or color process that will not fade or deteriorate under environmental conditions. These kinds of manufacturing problems can increase the cost of the product to the point that the color advantage is lost.

Although color has a strong influence upon market appeal, it has to be chosen skillfully. There is a cross-over point for many products where the cost of color styling cannot be justified. In general, the higher the sales appeal for color styling, the more costly are the techniques for achieving the finish. There is also a practical limit on how much extra the consumer is willing to pay to have a stylish product. The majority of us are willing to pay a little extra for a reasonable amount of styling. There is, however, a smaller group of people willing to pay considerably more, and who would never buy anything but an elegant and tastefully styled product. Of course, there is a small percentage of the market who will not be willing to pay any additional price for a stylish finish (just rust-proof it, and let it go at that). This is to say that although sales appeal is increased by styling, it is decreased by high cost. Thus the cost of increasing sales appeal through styling can lose out to the sales appeal of lower cost. The real problem lies in determining how much cost can be added for increased styling before the customer will rebel and accept a less stylish product with a lower price.

SO WHAT'S THE POINT?

The point is that good engineering alone is not enough in a highly competitive market—function is only part of the value attached to a product. If your product is going to stay alive, it is going to have to satisfy the aesthetic needs of the consumer. Or, to put it another way, human emotion is an important parameter in the design of any product that will be seen and used by people.

If the design is going to involve people during its use, those people are a realistic constraint upon the design problem. A good design engineer will not only be aware of all the psychological factors involved but will have the good sense to consult an expert appearance designer to aid in achieving a useful and satisfying product.

SELECTED REFERENCES

1. Appel, W. H. "Trends in Appearance and Design." *Product Engineering,* Sept. 21, 1959.
2. Birren, F. *Functional Color.* Crimson Press, 1937.
3. Bulinkis, J. J. "Industrial Design." *RCA Engineer,* Dec. 1975, Jan. 1976.
4. Gilbert, A. R. "Designing for Form as Well as Function." *ASME Paper,* 1973.
5. Ketcham, H. "Color in Design." *Machine Design,* Feb. 7, 1957.
6. Ketcham, H. *Color Planning for Business and Industry.* Harper & Row, 1958.
7. Koff, R. M. "Pretty—But Does It Work?" *Product Engineering,* Dec. 30, 1957.
8. Latham, R. S. & Siddall, J. N. "Aesthetics of Design." *Product Engineering,* Nov. 11, 1963.
9. Love, S. F. "Value in Design Takes Many Forms." *Design Engineering,* Dec. 1976.
10. Smart, W. O. "Industrial Design and the Engineer." *SAE Paper,* 1977.

9

Communication . . . Selling It

Engineers have two reasons for communicating: first, *to inform;* and second, *to persuade.* If they undertake a research project, the ultimate goal is to inform others of the new knowledge, facts, discoveries, techniques, etc. But, if the activity is a design project, the objective is to sell, gain support, solicit backing, influence skeptics, overcome resistance to change, convince people, and get action!

Design engineers produce ideas. All of the ingenuity and planning goes on in their heads. If engineers cannot successfully communicate their ideas to inform others of what they have in their minds—the entire effort has had it! The best idea in the world is a dead loss if no one can be persuaded to act upon it. All design engineers are faced with the task of presenting their ideas so that they will be understood. The immediate supervisor and everyone up-the-line in management must be sold on the idea and motivated to do something about it.

Thus, it is quite apparent that one of the most important jobs within the design process is communication and persuasion. At every decision-making stage throughout the entire design process, the project engineer has to gather all the available information, data, and ideas—and organize it in an intelligent and efficient way so that it can be understood and will generate a favorable response.

Since the success of every design effort depends heavily on good salesmanship, it is important, and indeed vital, that every engineer learn what it takes to communicate and persuade.

HOW TO WIN FRIENDS

Thomas Edison once remarked that society is never prepared to receive a new invention. Everything new is resisted. All of us basically fear major changes. They represent a threat to our security, our well being, our habit pattern, our comfort. People rise up against a new proposal, not because they dislike innovation, but because they fear the sociological effects that *might* come about.

All of us will bitterly resist a valuable idea, even if we recognize its good attributes, if we happen to be rubbed the wrong way or if it jeopardizes our security. We all have our own criteria for acceptance or rejection of a proposal. If someone misjudges what is important to us and fails to show us that the proposal fits our needs, he or she will never win our support.

Well, then, how do you win someone over? Here are some of the basic rules of good salesmanship:

> A young engineer from Saint Jo
> Wrote reports that were packed full of snow,
> Without an exception
> The frosty reception
> Left him standing alone in the cold!

- **Arouse their interest!** It's just as the old farmer claims—the only way you can get anywhere with a mule is to beat him over the head with a two-by-four until you get his attention. The first 10 words can be more effective than the next 10,000. Tell them immediately and simply what you have that you know is most important to them and they will be interested.
- **Watch the snow job!** If you lay down a big snow job you stand an excellent chance of freezing them out. Be direct and sincere. It takes very little talent to recognize a snow job.

If you are mistrusted, you are lost. To avoid being mistrusted, use only words and terms that are familiar to the receiver. Redefine technical information so that it has meaning in his or her frame of reference. Document your work from acceptable sources which are familiar to the person you are informing. Don't clutter your information with materials that have no direct relationship to the point you are making.

- **Be convincing!** Facts and information that are not obviously related to the situation are not convincing. Make the benefits clear and back them up with facts. Know what the people are looking for. Show exactly how your design fulfills these criteria. If they can't see that you have solved what they feel is the most important aspect of the problem, they won't be convinced.
- **Know your clients!** Find out what they are responsible for and what they consider when they make a decision. Each person in the management echelon who must contribute to the approval or rejection decision has a different set of criteria as a basis for reaching a judgment. You must tell them what they want to know or they won't be interested. It is extremely important that you know what they need, what they can understand, and the responsibility they have for the project.
- **Account for inertia!** All of us are hostile to a plan from the beginning until you win us over. We all inherently resist a proposal because we're afraid of what it might do to us. It's not the technical change that bothers us. It's the effect this change will have on us personally. If we have to be involved with it in any way, we want to make sure we don't get hurt. We are extremely protective of our status, our identity, our prestige, our job security, and our comfort.
- **Know what makes them buy!** The individuals you are trying to persuade will go along with the proposition only when they are satisfied that there is something in it for them. To put it that way makes the whole thing sound crass, but the point is that we all have basic desires, needs, and goals we want to satisfy. All of our decisions are weighed against things we regard as important to our involvement in the case. Indeed, there are a variety of personal motives behind our

acceptance of a proposition, but they generally fall into the categories of power, prestige, identification, loyalty to a group or cause, security, adventure, etc. Your success depends on knowing the people you are trying to sell. It will help you prepare your information for a good reception.

> Pity the day he got carried away
> And forthrightly was put on the shelf,
> He claimed reporting was long since passé
> And the project could speak for itself!

If you design, you have to sell! Someone has to be convinced that it's a good idea. And everyone who has to be convinced has to have a custom-made presentation. What will "sell" one person may antagonize another. The farther up the management ladder your selling campaign goes, the more important it is to be brief, concise, and less technical. High-level management is interested in product acceptance and cost. Lower management is interested in product practicality and the technical details of development and production.

If you are misunderstood, it is your fault.

Regardless of the level of concern, if you are misunderstood, it's your fault. If people object to your proposal because they have jumped to the wrong conclusion, you helped them jump. If they ignore your arguments and base their judgment on other criteria, you flubbed your presentation by not accounting for all the needs and concerns surrounding the problem and its acceptance.

The responsibility for selling is on the back of the salesman. The only reason one should fail is that the design is no good. If it is a good design idea and it never got off the ground, then there must have been a lousy job of selling the idea to the decision-makers.

THE LIVE PRESENTATION

The more important a design idea is, the more people need to be informed and influenced. In many companies, the project originator must make an oral presentation to a project-approval committee that is responsible for allocating funds and approving further development of the project. As the project progresses, more presentations are required. The project must be reviewed, and new groups of people must be informed as they become involved: such as tooling, production, purchasing, and testing groups.

Eventually, elaborate presentations must be made to higher management, the sales staff, and the distribution people, to educate them to the entire capability of the design. The engineer has the frequent and necessary job of preparing and delivering illustrated talks and demonstrations to keep the project alive and understood.

> There was a young fellow from Mission
> Who always attracted derision,
> Although he was bright,
> He ne'er saw the light
> And was lousy at verbal transmission!

Surveys of recent engineering graduates have revealed one thing above all others which engineers regard as their greatest post-graduate deficiency—not being able to make an effective presentation. Here are some helpful hints that will pay off when you have to prepare a "selling" talk:

- **Be brief!** This is one of the most important things you can do. If you take longer than 15 minutes to say what is needed, you are wasting everyone's time and losing your audience. Deliberate planning for a short talk forces you to eliminate the nonessential and talk about *only* that which is essential to your argument. Find out what your audience is basically interested in and make sure they get it in a clear, concise way!

ARE YOU GOING TO USE SLIDES?
Then, custom-make them for your presentation!

Do:

- Make special simplified sketches with double-weight lines.
- Use lettering that is twice normal size (¼" minimum height for 8½" x 11" format). Leave generous borders.
- Use color and shading to focus attention, provide interest, and relieve the monotony.

Don't:

- Don't clutter! Use only the essential information for the idea being illustrated. Leave generous amounts of space around the figure.
- Don't try to put a lot of points into one slide. Illustrate only one or two ideas per slide.
- Don't reproduce a figure from a text or manuscript for visual display unless it satisfies the above criteria. Usually text figures are not bold enough. They often contain too much information, and their type is often too small to be read by the audience.
- Never! Never read your slides to the audience. Talk about them; add information orally; point out interesting features; and let the audience do the rest.

- **Use visual aids!** And use good ones! A sloppy, poorly planned colorless illustration is just as detrimental to your presentation as being sloppily dressed and using bad grammar. If ever it pays to go to the trouble to do a good job, it is here. Sketches, diagrams, pictures, movies, charts, or models are almost a necessity in the presentation of technical design information.

There are available many visual aid devices such as overhead projectors, slide projectors, flip-charts, felt-boards, and movie projectors. The felt-board or flip-charts are useful only for presentations to small groups (not more than 15 people). The overhead projector and the slide projector are useful for larger audiences. The overhead projector can be used for sequence illustrations using overlays, and it allows you to make accenting marks and additions to the sketches as you talk. The new automatic slide projectors, however, are most widely accepted.

Organize your illustrations so that they can be automatically projected or can be changed by using an unobtrusive prearranged signal to the projector operator. Avoid if at all possible having to repeat endlessly, "May I have the next slide please."

The preparation of a well-planned and well-prepared sequence of slides or illustrations serves as an attention-holding device as well as an excellent set of lecture notes. You are released from having to read your talk or refer to your podium notes. It thus makes it easier for you to conduct a relaxed, informal, and dynamic presentation.

LISTEN TO YOURSELF!

Record a trial run of your talks occasionally to determine if you are cluttering your speech with redundant noises. Some typical redundancies to look for are:

Uh!
And ah!
In other words! (*or,* for instance!)
By and large!
And so forth! (*or,* et cetera!)
Don't you see? (*or,* isn't that right?)
All right!

We all tend to overwork a pet word or phrase, and there are few of us immune from the habit of saying "uh" or "ah" to fill up the silence while we're thinking of the next remark.

You'll never know you're guilty of these *irritating* little expressions unless you listen to yourself!

- **Be yourself!** Just get wrapped up in your subject and you have taken care of nearly all the Do's and Don'ts for making a good talk. The important thing to remember is that you are talking about your idea and your work; you know more about it than anyone else in the room. Talk as if you were conversing with a group of friends who have a great deal of interest in what you are doing—then let your natural enthusiasm do the rest.

FIVE POPULAR WAYS TO DIVERT YOUR AUDIENCE WHILE YOU TALK

- **Metronome!** This is one of the most popular: Stand first on one foot, then the other—shift your weight approximately five times per minute (as if you were in distress).
- **The Fig Leaf!** A popular stance—clasp your hands with arms hanging loosely in front of you.
- **The Double Grip!** Grasp the podium firmly on each side with arms locked stiffly (as if you expected the room to do a barrel roll any moment).
- **The Will Rogers!** Stand with all your weight on one foot, one hand in your pocket. Stare at the floor two feet in front of you.
- **The Barroom!** Put your weight on one foot with the other foot balanced on the toe, and lean one elbow on the podium.

- **Be well organized!** Pick out what is important, what the group needs to know to take action—describe it clearly and quickly, illustrate it, back it up with *pertinent* facts—and quit! Remember the first ten words can make or break you. The opening remark must be well planned to set the stage

and focus everyone's attention on what you intend to get across. Speak strongly and slowly, let them know you know what you're doing, and immediately focus their attention on your first illustration. Don't read the slide or describe the diagram. The audience isn't that dumb. Let them look at it while you point out the importance or significance of the idea being illustrated.

- **Use models!** Nothing convinces better than a working model. Model demonstrations, however, are best displayed using color movies. Even if you can get a live demonstration into the auditorium, it will be difficult for most of the audience to see. If you can, bring the model, display it for individual inspection after the talk, and do not use it during your presentation.
- **Watch your body language!** Most of us are unaware that the way we look, move, gesture, and behave sends out powerful messages to our audience. It has been reported that up to 80 percent of our communication is done with nonverbal messages–body language. And, surprisingly, the research shows that when the nonverbal message conflicts with or contradicts the verbal message, everyone will believe the nonverbal message first. If the way you are dressed or the mannerisms and behaviors you display are out of context with or contrary to what your audience expects or what they are hearing, you may very well be unselling rather than selling. If they see someone who looks bored, or scared, or nervous, or flippant, or insincere, they are apt to be suspicious of the verbal message–if they hear it at all. You must *appear* competent, enthusiastic, sincere, honest, and dedicated. It is a safe bet that if you truly *are* all these things your appearance and actions will convey it. Don't sabotage yourself by cluttering up your message with a lot of irrelevant and

> You can NOT
> NOT communicate!
>
> —
>
> Your looks will always
> Outshout your talking!

contradictory body-language "noise." Get someone to watch your presentation and give you an audit of your body language. Maybe they can help you clean it up. We all tend to develop habitual mannerisms and behavior styles that we have never noticed but that may cause our audience to be distracted and diverted from our intention.

- **When finished—sit down!** If you have completed your presentation and have briefly summarized the important points (tell-um what you told-um), then sit down! You don't have to make any remarks to let them know you are through. Don't get caught using trite, half-embarrassed remarks such as, "Well, that's all I have to say," or "That concludes my talk," or "I guess that wraps it up," or "In conclusion I would like to say . . ." or "Thank you very much." Don't thank your audience. They are not doing you a favor (even though you may feel as though they are).

ACTIONS SPEAK LOUDER THAN WORDS (10)

Warm behaviors that "turn them on" (positive response):

- frequent eye contact
- smiles
- nodding of head
- wide-open eyes
- raised eyebrows
- leaning forward
- intonation and inflection
- more voice volume
- higher speech rate
- hands held outward
- head raised
- body erect

Cold behaviors that "turn them off" (negative response):

- arms folded
- hands on hips
- looking away
- looking at ceiling
- frowns
- infrequent eye contact
- being over-precise
- noise words (ah, O.K.)
- high-pitched voice
- shoulder shrugging
- slower speaking rate
- shaking of head (negative)
- body slouching
- hands folded

Someone once remarked, "The world stands aside for people who act like they know what they are doing." You will gain a tremendous amount of respect, attention, and interest if you walk up to the podium and speak so that everyone senses that you know exactly what you're doing and why you're doing it. Your energy and your confidence will shine like a neon sign and it will have a vital influence on your audience. Be sincere, be earnest, but be yourself!

THE WRITTEN REPORT

The basic rules for a good written communication are simply:

- Tell them *what.*
- Tell them *why.*
- Tell them *how.*
- Tell them *"so what."*
- Tell them *briefly.*

The main reason for a written technical report is to tell people what they need to know in order to make a decision concerning the work done. They are not interested in a minutely detailed account of "The Perils of Pauline." They want to know whether you have achieved the answer to the problem or not. They want to know what the design will do and won't do and any other information that is pertinent to making a decision regarding subsequent action. They want the report to be brief, concise, and to the point.

The most important part of your report is the first page—the abstract. The abstract, or design summary, should be a one-page (or one-paragraph) summary of the *RESULTS* of the work. If you are describing a design, it should tell exactly what the design is capable of, what its important features are, and what ought to be done about it (why the reported information is important). Don't review what you did—summarize the *findings.*

This one-page synopsis helps the readers get all the important information quickly. It enables them to determine quickly if the report contains information in which they're interested—and whether they should take the time to read the whole report. It saves them from having to dig for the valuable and important information in the body of the report. A concise, informative, well-written summary will not only be appreciated by all your readers, it will have a considerable influence on their attitudes toward the entire report.

Those who want and need the information don't want to spend a lot of time getting it. If it takes them hours to puzzle out what you did, you are lost. They don't have that much time—if they did, they wouldn't need you to report to them. Here are some general rules for developing an efficient report:

- **Summarize, in brief,** the background of the work being reported, to show why it is important, needed, and valuable.
- **Tell the reader what you did,** without showing all the gory details. If you have used extensive calculations and technical resource material, include it in the appendix and merely outline major steps in the body of the report.
- **Use three-dimensional sketches** and diagrams for describing the physical aspects of the work. Summarize the results of the computations with simple graphs and diagrams. Wherever possible, use illustrations to save written words. (Don't catch yourself in the trap of spending two pages describing a figure that should speak for itself.)
- **Document the facts,** data, and technical information that were used by listing the sources and authorities. This is important in establishing confidence in your work, as well as making it possible for it to be checked and verified.
- **Leave out irrelevant information.** Show only what is necessary to state the case.
- **After it is all written—rewrite it!** When you do, try reducing the length of the report by at least 25 percent—and see how many of those "big shot" words you can eliminate.

WORDSMANSHIP

Good communication by the written word obviously is going to be influenced by the words you use and the way you lay them end to end. The cardinal rule is to use little ones, and little old everyday ones at that. There is a tendency for all of us to yield to the temptation to impress the troops by using a bunch of five dollar words (which we had to look up); scatter in some big-sounding technical "in language" terms; and use a lot of descriptive adjectives—to wit, to mount a big snow job.

There are some good rules to follow regarding the use of words. For example:

- **Use short sentences!** Everyone has a tendency to use long sentences when writing. Check your sentences for unnecessary phrases and superfluous adjectives. Break up compound sentences into simple ones. Professional writers average about 16 words per sentence; amateurs average more than 25.
- **Watch for multisyllable words!** Most words having more than three basic syllables (not counting endings like "ed" or "es") can be replaced by a simple word.

BIG WORDS VS. LITTLE WORDS

Don't say . . .	When you can say . . .
approximately	about
commence/initiate	start
endeavor	try
proceed	go
utilization	use
voluminous	large
modification	change
termination	end
contributed	gave
superfluous	extra/useless
subsequent	next
assistance	aid

- **Watch for redundancies!** Never repeat a word or phrase within a sentence or a paragraph if it can be avoided. We all have our pet expressions. It is easy to overwork them. Use variety in wording.
- **Don't use sweeping superlatives!** When you are trying to sell and motivate the reader, you can be drawn into the trap of using such words as largest, most, best, greatest, etc. By doing so, you only challenge the reader to doubt and to want to argue. Be realistic!
- **Avoid clichés and colloquial expressions!** You will do a better job of communicating if you use an easy, informal, conversational style. But, there is a limit to informality. Our everyday, conversational language is cluttered with slang, poor grammar, and quaint expressions that have little value in a good report. Remember, if your wording is odd or crude you will divert the reader from evaluating your report to evaluating you! Don't get corny, but don't be stuffy either!

CLICHÉSMANSHIP

Don't write . . .	When you mean . . .
Has not got any . . .	has no
In the great majority of the cases . . .	most often
In regard to . . .	?
A major consideration . . .	important
There is a dearth of information . . .	little is known
To a large extent . . .	much of
Entail considerable expenditures . . .	costly
In the vicinity of . . .	near
By and large . . .	?
A considerable amount . . .	some/much
In the main . . .	basically/usually
In other words . . .	or
On the other hand . . .	?
As shown in Figure 4 . . ./ Fig. 4 is a picture of . . .	Figure 4 shows
At the present time . . .	now
As a matter of fact . . .	?
It has been brought to our attention . . .	I'm told

- **Watch spelling and grammar!** Engineers can't spell. Admit it, and use a dictionary or get someone to check your spelling! Not being able to spell is one thing, but letting someone else find out about it is unforgivable—especially when you are writing a report that represents your ability.
- **Check your Fog Index!** Just as you should record your oral presentations once in a while, so should you sample your writing to get some idea of its clarity. The Fog Index (see box) will give you a good indication of how much confusion you are creating. It not only pays to be brief, but it pays to be easily understood. According to Gunning (7, 8) if you have a Fog Index above 12, you are in trouble. Most good literature has an index less than 8. The Index is an approximate indication of the number of years of schooling a person should have to understand your writing without difficulty.
- **Edit and rewrite!** Just as any engineering has to be reworked and optimized, so does your writing. You are a rare person indeed if your first draft is refined and polished. It is surprising how many beginners fail to see the need for editing and rework. Don't be reluctant to cast out words and sen-

tences and paragraphs and do them over. Edit to reduce length and to clarify! And do it over more than once.

GUNNING'S FOG INDEX* (7, 8)
(The number of years of schooling a person would require to read a passage with ease and understanding.)

- Pick a sample of text 100 words long. Find the average number of words per sentence by dividing the total number of words by the number of sentences.
- Count the number of words of three syllables or more.
  Don't count:
  Words that are capitalized.
  Combinations of short words (like "bookkeeper," etc.).
  Verbs that are made three syllables by adding "ed" or "es" (like created).
- Add the two factors above and multiply by 0.4. This is the Fog Index.

If the Fog Index exceeds 12, the passage should be rewritten! A well-written report should have a Fog Index of less than 8 to come through loud and clear.

WHEN YOU ILLUSTRATE YOUR REPORT

Take time to make your figures and graphs interesting and informative. Sketches, photographs, and diagrams are just as important as the text, and deserve as much attention. Although illustrations are valuable and attention-getting, you need to exercise some restraint. Don't clutter your report with figures that have no useful purpose. Good illustrations are the most efficient means you can use to convey information. Use them to save words.

There are several types of illustrations that are commonly used in engineering reports. Here are some useful tips for using them effectively:

- **Sketches**—Whenever possible use three-dimensional perspectives with a modest amount of shading for depth and accent. The use of exploded assembly views, cutaways, and phantom lines transmits a lot of information.

*Method of computing the Fog Index is drawn from *More Effective Writing in Business and Industry* by Robert Gunning, published by the Industrial Education Institute of Boston. This copyrighted material is used with written permission of the author.

- **Photographs**—If you are illustrating hardware and machine setups, pay particular attention that the dark metallic parts are adequately lighted and have a contrasting background.
- **Diagrams and schematics**—These are the most common forms of report illustrations. They are simple, clean, and can be easily read. Use *heavy* outlines contrasted with light secondary lines. Don't clutter the diagram with unimportant detail, use large readable type, and leave generous space around the diagram.

GRAPHS

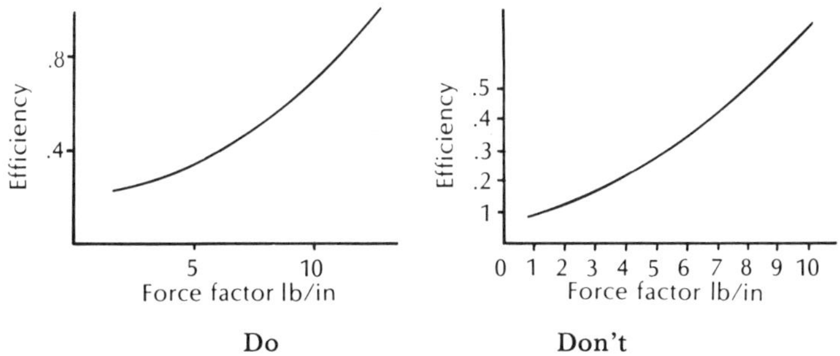

(1) Do not clutter up the coordinate dimensions with numbers too closely spaced. (Minimum ¾"–1" apart.)
(2) Use scale divisions of approximately the same spacing on both ordinate and abscissa. (Minimum ½" apart.)

- **Graphs**—There are two reasons for presenting graphs: to show a trend due to the variation of a parameter or the relative trends of two parameters, and to present data. If you are displaying trends or parameter variations, it is not necessary to clutter the graph with grid lines. If you are plotting data that you expect the reader to determine from the graph, show the major grid lines.

The trace is a plot of computed information. Show it as a smooth curve without displaying the plotted curves. However, if the trace is showing the trend of experimentally determined data, show each plotted point as a small circle. Draw the trace to but not through the circles. **Never** draw a freehand trace—use a French curve and draw a smooth, bold, continuous curve.

Do not crowd the numbers on the scales. A good rule to follow is to allow a minimum of ½" spacing for all grid numbers. Always indicate the name, symbol, and units of the parameters related to each scale on the grid.

- **Tables**—Engineers in general seem to be sold on the use of tables to display the results of their work. However, tables have some definite limitations as efficient communication devices. Unless the tabulated information is to be specifically used and referred to in the text, or is to be a reference source for the reader, it has little value. If you are interested in showing trends within the data you should plot it on a graph. It is extremely difficult to see meaningful trends in columns of tabulated numbers. A good rule-of-thumb is to avoid using tables. They are difficult and expensive to prepare. Unless the information is intended to be a reference source for data that cannot be accurately read from a graph, there is no other useful justification.
- **Color**—Use it! It is powerful and attention-getting when used to accent and clarify your drawings and sketches. If the report must be duplicated, the use of colors may not be practical. But for single "in house" reports, or a plush "slick" proposal, color can hardly be ignored for effective communication.

The use of sketches and drawings is "a must" for the design engineer. Engineers who cannot make a clear and meaningful drawing of their ideas are severely handicapped. This cannot be passed off by assuming that you can always find someone else to do it for you—how are you going to tell that person what to draw?

You'll rarely ever see engineers conversing without sketch pads and pencils. Learning to draw good sketches is as important to an engineer as learning to write. If you are guilty of sloppy unintelligible sketches, you cannot take refuge in the excuse that you don't have any artistic talent. You learned to write without being a noted author. You can learn to sketch just as effectively as you can learn to print or write.

SO . . .

Communication is one of the most critical requirements in the entire design process. Everything is lost if communication between engineers fails. Engineering design is a team project requiring the coordination of the efforts and decisions of many engineers who are responsible for many diverse facets of the project. Each has to understand what the other thinks and does. Each has to develop ideas around the information of the others. Each has to make decisions based on the information received from the others. Most of the failures that occur can be blamed on a boggled exchange of information or on misunderstandings.

So! You can work like a dog—do a fine job of thinking out the problem—develop a valuable solution—and then you can blow it by a lousy presentation. A generator is useless without a good transmission system.

SELECTED REFERENCES

1. Cadrecha, M. A. "Take The 'Fog' Out of Your Writing." *Consulting Engineer,* Oct. 1964.
2. Carnegie, D. *How to Develop Self-Confidence and Influence People by Public Speaking.* Simon & Schuster, 1969.
3. *Communications* (Reprints from *Product Engineering*). McGraw-Hill, 1965.
4. Cooper, J. *How to Communicate Policies and Instructions.* Bureau of National Affairs, 1960.
5. Ellenwood, J. L. *So You're Going to Make a Speech.* Scribner's, 1951.
6. Foley, J. J. "Freehand Sketches—Here's How." *Product Engineering,* Feb. 9, 1959.
7. Gunning, R. "How to Improve Your Writing." *General Electric Review,* Nov. 3, 1952.
8. Gunning, R. *More Effective Writing in Business and Industry.* Industrial Education Institute, 1962.
9. Kirkpatrick, T. W., & Breeze, M. H. *Better English for Technical Authors.* Macmillan, 1961.
10. Knapp, M. L. *Nonverbal Communication in Human Interaction.* Holt, Rinehart & Winston, 1978.
11. Leathers, D. G. *Nonverbal Communication Systems.* Allyn & Bacon, 1978.
12. McCartner, T. O. "Perspective Grids." *Product Engineering,* Jan. 11, 1960.
13. McKnight, A. N. "How to Sell Your Idea." *Product Engineering,* May 11, 1964.
14. Pace, R. W., Peterson, B. D., & Burnett, M. D. *Techniques for Effective Communication.* Addison-Wesley, 1979.

15. Rosenstein, A. B., Rathbone, R. R., & Schneerer, W. F. *Engineering Communications.* Prentice-Hall, 1964.
16. Strunk, W. *The Elements of Style.* Macmillan, 1972.
17. Watztawick, P., Weakland, J., & Fisch, R. *Change: Principles of Problem Formation and Problem Resolution.* Norton, 1974.
18. Woelfe, R. M. "Better Report Writing Pays." *National Electronics Conference Proceedings,* 1962.

Index